THE GENE MASTERS

THE
GENE
MASTERS

*How a New Breed of
Scientific Entrepreneurs
Raced for the
Biggest Prize in Biology*

INGRID
WICKELGREN

TIMES BOOKS

Henry Holt and Company   ·   New York

Times Books

Henry Holt and Company, LLC
Publishers since 1866
115 West 18th Street
New York, New York 10011

Henry Holt® is a registered trademark
of Henry Holt and Company, LLC.

Library of Congress Cataloging-in-Publication Data

Wickelgren, Ingrid.
    The gene masters : how a new breed of scientific entrepreneurs
raced for the biggest prize in biology / Ingrid Wickelgren.—1st ed.
        p. cm.
    Includes bibliographical references and index.
    ISBN: 0-8050-7174-1 (hbk.)
        1. Human Genome Project. 2. Human gene mapping. 3. Geneticists—
United States 4. Genetic engineering industry—United States. I. Title

QH431 .W483 2002
611'.01816—dc21   2002067321

Henry Holt books are available for special promotions
and premiums. For details contact: Director, Special Markets.

First Edition 2002

Designed by Victoria Hartman

Printed in the United States of America
1   3   5   7   9   10   8   6   4   2

TO BOB AND LARS

Contents

THE GENE MASTERS

PROLOGUE: UNFINISHED BUSINESS

On October 4, 1994, Francis Collins sat impatiently in a hotel ball-room waiting for the U.S. Secretary of Health and Human Services, Donna Shalala, to finish her speech at a Washington, D.C., genetics conference. The towering, lean human geneticist was anticipating the urgent, invitation-only meeting upstairs to take place immediately afterward. The previous year, Collins had assumed the leadership of the largest-scale biological effort the U.S. National Institutes of Health (NIH) had ever undertaken: a fifteen-year project to decipher the genetic blueprint of a human being. Now, Collins faced perhaps the biggest challenge of his tenure so far.

A few years before Collins assumed his post, one of the NIH's star scientists, the cowboy biologist J. Craig Venter, had invented a clever method of finding genes, the most informative segments of the long DNA molecule that is the human genetic blueprint. But his techniques were booed so roundly by jealous and shortsighted colleagues that he'd been essentially shoved out of academia and into the private sector. As a result of his defection, a large and enormously valuable database developed using Venter's revolutionary method had fallen into private hands. Though they had tossed out the seed from which this biological bounty sprouted, scientists were suddenly furious that this storehouse of knowledge was now largely inaccessible to them.

The database threatened to make the whole genome project irrelevant. Venter's business partner, William Haseltine, whose company Human Genome Sciences owned rights to commercialize Venter's work, claimed that their database already included the vast majority of human genes. If he was right, one company could own plum portions of the human genome even before Collins's researchers began decoding it in any serious way. But the more immediate problem was finding a way to allow researchers public access to this vital storehouse of knowledge, and prevent it from being locked up in a private database forever.

When the secretary finished, Collins sprang from his seat, dashed upstairs, and ducked into the small meeting room where the closed-door session was about to take place. Two dozen other top academic and industrial gene scientists also filed into the room, took their seats, and promptly began debating how to handle the situation.[1]

Almost none of the academics in the room liked the deal that Haseltine, who had somehow managed to get invited to the gathering, had offered them. Haseltine had said academics could see his and Venter's data as long as his firm received first option on commercial rights to any genes found as a result—and one to two months' notice on any resulting publications. But scientists found the notion of their research exclusively abetting a private company abhorrent; neither did they want strings attached to the publication of their work.

Collins agreed. Signing restrictive data-sharing agreements was totally unacceptable to him. He favored a second option: Work around the company and create a copycat database for academics. But that alternative was also problematic. Allotting federal funds for that purpose would be seen as a deliberate attempt to damage a successful company, putting Collins in hot water with Congress. Collins's predecessor, James Watson, had been pushed out of office in part because of actions perceived as harmful to private enterprise. Collins had spent months thinking about what to do, and then an answer had appeared out of nowhere.

Alan Williamson and Keith Elliston from Merck & Co. had swung into his office one day to say that the pharmaceutical giant was willing to finance the creation of a database similar to the one Haseltine was hawking. Merck wanted this basic genetic knowledge to be widely available, because it believed that would bring the greatest scientific progress. Of course, the deal would give Merck access to the same kind of data that its competitor, SmithKline Beecham (later Glaxo-SmithKline), the corporate partner of Haseltine's firm, was already tapping for its drug-research programs. And Merck really needed that data for its fledgling gene research effort.

Collins couldn't believe his good fortune. One of the world's most prestigious drug companies had just volunteered to solve his problem and pay for it with virtually no strings. Collins suggested the funds might go to a team at Washington University in St. Louis, which was best equipped to carry out the main work to build such a database.

But the support of the scientific community was crucial for anything to happen. Merck's flamboyant chief executive, Roy Vagelos, had made it abundantly clear to Williamson and Elliston that there would be no deal if there were any chance it would embarrass Merck and tarnish its pristine image on Wall Street. Although *The Wall Street Journal* had broken the news of the unfolding secret plan a week before, the October gathering was the first chance that Williamson and Elliston had to gauge academics' reaction to it. Would they embrace a genetic database paid for and orchestrated by a corporate giant?

Williamson and Elliston stood up in front of the group and described the details of the arrangement publicly for the first time. Merck was ready to go ahead, Williamson said from the podium. They'd lined up the Washington University team, which would be using expensive machinery to spell out the chemical letters in hundreds of thousands of small strips of human DNA. These genetic words would be deposited into a national data bank in Bethesda, Maryland, that anyone could access via the Internet. No encumbrances. No fee.

The announcement provoked strong reactions. From his perch in the front row, Haseltine, dressed as always in a tailored dark suit, was defiant. "You're destroying intellectual property!" he told the men from Merck. He accused the big drug firm of using its pocketbook to maintain its dominance and squash smaller firms like his.

Some of the academics in the packed room were also skeptical of Merck's motivations. There had to be a catch. This was just too good to be true. But Elliston calmly reassured them that Merck was not up to anything sneaky. The big drug firm's executives simply felt that such basic genetic data was precompetitive and ought to be widely disseminated. And Collins argued that academics really needed a completely unrestricted database to use the information for the variety of purposes that they had in mind. The academic researchers slowly began to see Merck as their white knight, saving them from Haseltine and Venter's attempt to enslave basic genetic data in weighty corporate chains. By the end of the meeting, it was looking as if Merck's plan might actually happen.

It was not yet a done deal, however. The details of the agreement had yet to be worked out, and one potential snag remained. Columbia University, one of whose researchers would have to supply the genetic materials to the Washington University team, was already in negotiations to license those materials to another company. Columbia lawyers feared a backlash from that company if Columbia agreed to simply give the materials away. The problem threatened to derail the project.[2]

Behind the scenes, Collins weighed in, emphasizing to Columbia that the undertaking was critical to the country. Other academics also lobbied Columbia on behalf of Merck. Finally, Columbia agreed to an arrangement in which the materials would bypass Merck and travel only among the academic institutions doing the work. Weeks after everyone else had signed on, Columbia said it would, too.

When the effort got under way the following year, Collins felt triumphant and relieved. He had helped broker a solution that undid the damage done by the private sector. Under Watson, the genome project had been too slow to recognize the importance of Venter's gene-hunting method and, as a consequence, had almost let a vital tool for extracting the gold from the human genetic code slip from their hands.

The Merck deal turned everything around, Collins thought. He figured that the battle between the public and the private sector—and between himself and Venter—over decoding genetic information was over.

But it had only just begun.

1

ENGINEERS PLANT A ROSE

The Human Genome Project was unleashed upon the scientific scene in the mid-1980s amid great tumult and uncertainty. It was born to admiring parents who were deemed unworthy, and then thrust into the arms of guardians who did not want it, or at least not as it had originally been conceived.

This project was the culmination of a century of research into the human genetic code, a lengthy molecular treatise at the center of every cell in the human body that describes in detail how to construct a human being. The treatise contains pages that determine the color of our hair and eyes, influence our ability to run a six-minute mile, and tell us our chances of becoming a mathematician or musician.

It holds darker secrets as well: tiny typos that can increase a person's odds of developing ailments like Alzheimer's disease, heart failure, diabetes, or cancer. Pinpointing these typos in the treatise could give scientists crucial clues to diagnosing, preventing, and treating the thousands of illnesses afflicting mankind. A detailed understanding of the genetic code could help researchers solve the mysteries of how the body develops and ages, and the evolutionary path that brought humanity to this Earth.

The first solid hints of our inherited secrets appeared in the mid-nineteenth century, courtesy of an Austrian monk named Gregor

Mendel. Mendel deduced the laws of inheritance by studying pea plants in the monastery garden. Observing how characteristics such as height or seed shape were passed from one generation to the next, he postulated the existence of invisible "elements" hidden inside the plants that parent plants passed to their offspring and that governed the plants' visible features. He came up with rules for how different combinations of elements would produce recessive or dominant traits.

Nobody believed Mendel at the time, but his work was rediscovered in 1900 and soon became the foundation for modern genetics. The nature of Mendel's elements—genes in modern parlance—was shrouded in mystery for decades. For much of the early twentieth century, most scientists thought that genes were embedded inside special types of proteins. Then in the 1940s, evidence slowly built that a mysterious molecule called DNA was in fact the Rosetta stone of inheritance.[1] But no one knew how DNA might work, or even what this molecule looked like.

In a stunning breakthrough, an American named James Watson and an Englishman named Francis Crick solved these riddles in 1953. They revealed the structure of DNA, in which two DNA strands twist around each other in a so-called double helix. Each DNA strand consists of a sequence of four different molecular components called bases, known by their first initials: A, T, C, and G—for adenine, thymine, cytosine, and guanine. Watson and Crick discovered that these bases clasped each other in pairs—A's always hugged T's, and C's hung on to G's—forming rungs that fit inside a molecular scaffolding resembling a spiral staircase.

Inside every human cell, separate spiral DNA staircases comprise each of forty-six chromosomes. Those chromosomes couple up as twenty-three pairs, with one pair mismatched in males. In men, the twenty-third pair consists of a so-called X and Y chromosome, making twenty-four unique types of DNA staircases in humans. If one stacked all twenty-four unique staircases on top of one another, one would produce a long ascent of some three billion steps. Each step would be a pair of Watson and Crick's hugging chemical letters, a so-called base pair. Mendel's elements, or genes, were scattered flights of stairs broken up by less critical steps.

Watson and Crick's discovery, for which they received a Nobel Prize in 1962, showed the world for the first time the true nature of the slate on which our genetic code is written. But researchers needed to understand how cells used this information. Soon they discovered that cells translate the functional segments of this DNA molecule, the genes, into proteins. Genes were blueprints for proteins, the workhorses of every cell and the machinery of life.

And so, one by one, biologists tried to unearth and isolate key proteins and the genes from which they were made. It was a laborious process, in many cases requiring years of effort to find a single protein cog in a cellular machine. And once a protein was found, researchers had no good way of producing more of it.

That changed in the early 1970s. In 1973, two California biologists, Herbert Boyer and Stanley Cohen, started the biotech revolution by sticking a piece of frog DNA inside a lowly bacterium. The bacterium acted as a tiny factory for the inserted gene, copying it along with its own genes every time it divided. In this and other experiments involving bacterial genes, the two scientists pioneered the field of genetic engineering, the ability to clone genes inside bacteria. (Gene cloning, or creating many exact copies of single genes, is not the same as "Dolly" the lamb cloning, in which an animal's entire set of genes is used to create another genetically identical animal.)

Two years later, Robert Swanson, a Silicon Valley venture capitalist, approached Boyer with a vision for creating a new class of drugs: Use the new bacteria factories to produce proteins that treat disease. On April 7, 1976, Swanson and Boyer founded the San Francisco–based firm Genentech, and the multibillion-dollar biotechnology industry was born.

Soon Genentech scientists cloned human insulin, which diabetics need to take to regulate their sugar metabolism. Bioengineered human insulin, the first recombinant DNA drug sold, eliminated the expensive process of extracting insulin from pigs and cows.[2] In 1979, Genentech researchers cloned the gene for human growth hormone, which some children need to take to reach normal height. Bioengineered growth hormone promised to be much cheaper and safer than the old form of the hormone, which had to be isolated from human cadavers and sometimes contained fatal diseases.

Despite the breakthrough of biotechnology, researchers still couldn't directly locate new genes. They had to work backward from a protein they had isolated and characterized first. But then in the early 1980s, biologists invented a way of finding important genes by their placement in the genetic blueprint. They studied extended families to identify tiny landmarks near disease genes that are passed down from generation to generation. In such a way, they pinpointed genes underlying inherited diseases such as Huntington's disease, muscular dystrophy, and, later, cystic fibrosis.

By the mid-1980s, biologists had found the approximate location of more than seven hundred human genes but had precisely identified only a dozen or fewer. At this rate, it would take centuries to find the rest of the estimated fifty thousand to one hundred thousand human genes.

So an obscure biologist proposed a preposterous plan that would usher the study of our genetic code into a new age. Instead of studying one gene at a time, he suggested finding them all using an emerging technique that enabled scientists to directly decipher the chemical letters of DNA molecules.

The Human Genome Project was the brainchild of a thin, dark-haired biologist named Charles DeLisi. In 1985, DeLisi arrived at the Germantown, Maryland, offices of the Department of Energy (DOE) to direct the agency's Life Sciences Division, then called the Office of Health and Environmental Research. DeLisi was a biologist with an unusual bent. At the National Institutes of Health, where DeLisi had spent a decade, he had developed an odd penchant for using computers to sift meaning from the human genetic code. DeLisi was fascinated with the information embedded in the sequences of A's, T's, C's, and G's that link together to form DNA and compose that code.

Though not known for its biological research, the DOE had long sponsored biology projects. In its former guise as the Atomic Energy Commission, it was the largest patron of genetics research in the United States, paying for studies of the health effects of ionizing radiation like that from the atomic bomb. By the 1980s, the agency's mission had broadened to include the health impacts of nonnuclear sources of energy. Agency biologists were keenly interested in new ways to understand the human genetic code, because that was the molecular slate on which toxins and radiation made their mark.

In October 1985, a document titled "Technologies for Detecting Heritable Mutations in Human Beings" landed on DeLisi's desk. It was a report from a meeting held the previous December in Alta, Utah, a tiny mountain enclave known for its ski resort. The meeting was called to discuss new methods for detecting DNA mutations, tiny changes in the DNA code that can be produced by environmental chemicals or radiation.

As DeLisi scanned the Alta report, his eye caught an offhand comment directed at the question: How do you determine whether genetic mutations are passed on from a parent to his or her offspring? The organizers of the Alta meeting had been particularly interested in the possibility of detecting a rise in mutations among survivors of the bombings of Hiroshima and Nagasaki and their children.[3]

In principle, the report read, you could compare a child's complete set of genetic instructions, or genome, with the genomes of his or her parents and pinpoint the differences. Taken literally, DeLisi knew it was a crazy idea. The human genetic instruction book was written as a sequence of approximately three billion of DNA's four chemical letters. Nobody would spell out all those letters just to find a few minute differences between a mother and daughter or father and son. The amount of effort would be overkill, and it might not work anyway.

But DeLisi was intrigued by the notion of spelling out the entire sequence of chemical letters that compose a human's genetic code. We could do a thousand other things if we could read that sequence, he thought. Who knows what would be discovered? Even developing the technology for such an effort would bring fabulous new findings, he figured.

The sequence of the human genome governed the workings of every cell in the body. Knowing the sequence of all those chemical letters, DeLisi realized, could help biologists find all the genes and begin to understand the functions of the proteins they produced, shedding light on bodily processes from etching a memory in the brain to the beating of the heart. From there, scientists could pinpoint mutations in genes—whether inherited or acquired—that cause, or raise the risk of, disease, an important step in understanding how to diagnose, prevent, and treat the plagues of humanity.

DeLisi's mind raced. He envisioned his idea capturing the imaginations of congressmen, his former friends at the NIH, and citizens across America.

DeLisi knew his plan was theoretically feasible. Genes, measuring a few billionths of a meter, are far too small to be seen. A microscope cannot find them. But a decade before, an Englishman named Frederick Sanger of Cambridge University and, separately, Walter Gilbert, a Harvard biologist, developed ingenious chemical tricks for determining the order of DNA's tiny molecular building blocks.

DeLisi went straight to the head scientific adviser for his division, Mortimer Mendelsohn, to share his inspiration. But Mendelsohn had deflating news. The idea had already been proposed and rejected. A year before, the chancellor at the University of California, Santa Cruz, had wanted to build an institute on his campus to decipher the human genome to boost the visibility of the university. He had been inspired by the big physics and astronomy projects of the day, and he had wanted to pioneer a biology counterpart. But top biologists concluded that the technology for the job was not yet available.[4] They proposed a more conservative plan, but even that fizzled for lack of funding.[5] The National Institutes of Health had not been interested. It was accustomed to funding small projects proposed by individual scientists rather than large, centrally organized, technology-driven endeavors.

DeLisi was not discouraged. He realized that the DOE was far more likely to support his plan than the NIH. In contrast to the health institutes, the Department of Energy was a place where biologists worked among engineers and physicists, where large-scale projects were par for the course, and technology was king.

DeLisi's organization had already been laying a lot of technological groundwork for such a project. Researchers at the DOE's Los Alamos National Laboratories had established an international computer database, called GenBank, containing all the known genetic sequences. Los

Alamos biologists had also been building a large library of chunks of human DNA, initially as a resource for smaller-scale genetics projects.[6] But this library could conceivably be adapted to supply the fodder, the necessary bits of DNA, for sequencing the human genome.

So DeLisi was optimistic. There was clearly something to build upon.

He decided to take the scientific community's pulse on this issue. He organized a workshop in Santa Fe in March 1986 a few miles from the DOE's Los Alamos lab.[7] In addition to some sixty academic scientists, DeLisi invited representatives from all the major biology funding agencies, including the NIH and Environmental Protection Agency (EPA). But none of the agency officials showed up. The NIH was outright hostile to the idea. It was not eager to support technology development. "No one was interested," DeLisi later said. "It's unbelievable. This is so mainstream now; people don't realize how radical it was."

A few prominent biologists did grasp DeLisi's vision, however. Harvard's Walter Gilbert, the sequencing pioneer, declared at the start of the meeting: "The total human sequence is the grail of human genetics."[8] Others echoed Gilbert's enthusiasm, including Charles Cantor, a Columbia University biologist, and Leroy Hood, of the California Institute of Technology, who would soon unveil an automated technology for sequencing DNA.[9]

The Santa Fe attendees agreed that genome sequencing was a worthy goal, but thought it couldn't begin immediately. For one thing, sequencing needed to become cheaper and faster. Sequencing had to be done by hand then. State-of-the-art laboratories could sequence only about five hundred bases in a twenty-four-hour period.[10] In practice, though, the average sequencing rate in most labs was less than fifty thousand base pairs *per year.*[11] At that rate, it would take one hundred labs six hundred years to cover every base in the human genome once. Sequencing was also prohibitively expensive. At the current cost of $5 to $10 per base pair, sequencing the entire human genome could cost as much as $30 billion.

The other problem: The methods for determining the chemical sequence of DNA could not read an entire human chromosome whole. The twenty-four unique human chromosomes, which each contained a separate segment of the human genetic code, were tens of millions of base pairs long. But sequencing technology could read only a few hundred base pairs at once. Thus, researchers needed to produce a collection of tens of thousands of short pieces of human DNA that could be read and later put back together to re-create each human chromosome. Such a collection didn't yet exist.

For simple organisms, such as viruses, one could just collect many copies of a virus's genome (from so many viruses) and chop up each one randomly into tiny pieces. That would generate many tiny overlapping pieces of DNA that could be individually sequenced and then, by detecting their regions of overlap, reassembled to re-create the genome. But the Santa Fe attendees decided that strategy would not work for the human genome. There would be too many pieces. A human brain certainly couldn't fathom putting them back together, and computer power was not yet up to this task either.

Instead, the scientists settled on a two-step procedure. They would chop each human chromosome into relatively large chunks and figure out how those chunks fit together to re-create the chromosomes. In step two, each chunk, or part, would be fragmented into the smaller pieces that people could sequence, and those sequences would be reassembled to re-create the chunk.

The technique was akin to reading a Bible by shredding it one page at a time, reading each of the resulting shreds, and then reassembling each page from the shreds before putting the pages back in order according to their page numbers. In the end, one copy of the Bible—or human genome—would be whole again, and completely deciphered.

But this two-step method of sequencing also required technology that wasn't yet available. It required a way of copying large pieces of human DNA that would reliably preserve their original sequences and a so-called physical map representing the order of the chunks.

At the meeting's close, DeLisi asked all the attendees to pen their private conclusions about how to proceed and to send these to him. Back in his office in Germantown, DeLisi promptly read the dozens of letters he received from the scientists and summarized the findings in a memo he sent to his boss, the director of the Office of Energy Research.

DeLisi trumpeted the proposed project as a way of spurring the development of technologies critical for identifying genes that lead to disease. That, in turn, could help spawn important new diagnostic instruments and ultimately therapies. For the first five years, researchers would work on the physical map of the genome and invent technologies that would make sequencing cost effective. After that, researchers would begin rapidly sequencing the genome. It was a fifteen-year proposal, to start in 1986 and finish in 2001.

His boss gave DeLisi $5 million to get started. But most NIH-funded biologists wanted nothing to do with DeLisi's project. At a June symposium at Cold Spring Harbor Laboratory on Long Island, Harvard's Gilbert stood before a crowd of three hundred to advocate a dedicated effort to sequence the genome. He estimated that this might require thirty thousand man-years of work—say, a thousand scientists working for thirty years. Then he wrote "$3 billion" on the board. That was his estimated cost for getting the job done, and it unleashed an uproar from the audience. To biologists, for whom a $1 million grant was large, Gilbert's figure was astounding—and threatening.[12]

An influential MIT geneticist named David Botstein stood up at the meeting and said, "If it means changing the structure of science in such way as to indenture all of us, especially the young people, to this enormous thing like the Space Shuttle, instead of what you feel like doing . . . we should be very careful."[13] He derided the notion of DNA sequencing, noting that if Lewis and Clark had followed a similar approach to mapping the American West, a millimeter at a time, they would still be somewhere in North Dakota. The audience applauded.

Later that summer, a letter written to the scientific journal *Nature* argued that "sequencing the genome would be about as useful as translating the complete works of Shakespeare into cuneiform, but not quite as feasible or as easy to interpret."[14] Another MIT scientist, Robert Weinberg, scoffed: "I'm surprised consenting adults have been caught in public talking about it."[15] And an English biologist named Sydney Brenner quipped that the sequencing job ought to be parceled out to prisoners, with the most work given to the worst offenders.[16]

Like Brenner, most traditional biologists regarded sequencing DNA on a massive scale as mindless and dull. Deciphering a long, monotonous string of A's, T's, C's, and G's was nothing more than a giant fishing expedition. They didn't see the value in it. They felt that answering narrower biomedical questions, such as efforts to find specific genes, would be far more interesting and useful. Those at least gave meaning to the molecules, tying them to a function in a cell or a role in a disease.

Others felt sequencing the genome was wasteful for another reason: At least 90 percent of the human genome was thought to have no biological function, containing neither genes nor sequences for controlling genes. It was so-called junk DNA. Thus, even if people could interpret the functional parts of the genome, determining the vast bulk of the sequence would be a waste of time.

Behind a lot of the scientific criticism was fear. Biologists worried that such a big project would squash their independence and the cottage-industry culture of biology in the process. Historically, biology projects were small-scale endeavors, involving a handful of graduate students, postdoctoral fellows, and other lab staff. They involved testing interesting hypotheses. They were not large-scale, centrally organized, technology-driven efforts.

By contrast, DeLisi was inspired by the idea of bringing high technology to bear in a creative way to advance the field of biology. Automated sequencing machines were on the horizon, and DeLisi could easily see how computer software would help scientists make sense of the sequence, too. Others at the DOE, fundamentally a physics organization, also relished the technological challenges involved.

In April 1987, a DOE advisory panel recommended that the agency lead the nation in launching "a large, multi-year, multidisciplinary, technological undertaking to order and sequence the human genome." Mendelsohn wrote in his report that the Human Genome Project Initiative should transform the activity of reading human genes from "an inefficient, one-gene-at-a-time, single laboratory effort into a coordinated, worldwide, comprehensive reading of 'the book of man.'"

The idea that DOE should lead this effort, Mendelsohn acknowledged, might seem audacious. But he argued that the DOE alone "has the background, structure and style necessary to coordinate" such a program, citing the necessity of "strong central control," and DOE's expertise in managing complex, multidisciplinary projects.[17]

But the NIH director at the time, James B. Wyngaarden, wasn't about to let the DOE, of all institutions, horn in on what he regarded as the NIH's turf. Wyngaarden first heard about DeLisi's proposal at a conference in London in June 1986, when somebody asked him what he thought of the DOE's plan to spend some $3 billion to sequence the human genome. Wyngaarden was shocked. The NIH had for years been systematically building up a bank of genetic information related to human disease. DeLisi's proposed genome project had all the earmarks of a hostile takeover.

Wyngaarden asked the directors of the NIH's institutes, such as the National Cancer Institute and the National Institute of Neurological Disorders and Stroke, what they thought. All but one of them had no interest in being part of it. They regarded a dedicated effort to decipher the human genome as an invasion of their disease-specific territories. They figured that the genome would be decoded eventually anyway, so why bother with a big centralized undertaking?

But Wyngaarden was determined to steal the project away from the DOE one way or another. To build scientific support, he commissioned the National Academy of Sciences, the nation's premier scientific body, to evaluate the genome plan. The academy convened an all-star review team including James Watson, Harvard's Gilbert, and Bruce Alberts, a biochemist at the University of California, San Francisco. The panel spent more than a year gathering information from experts in various genome subspecialties to decide on the best course of action.

As the committee deliberated, biologists continued to fight the DOE plan. The Council of the American Society of Biochemistry and Molecular Biology weighed in in June 1987. "The Council wants to state in the clearest possible terms our opposition to any [centralized project] . . . to sequence the human genome, " the statement read in part. "It is of the utmost importance that traditions of peer-reviewed research . . . not be adversely affected by efforts to map and sequence the human genome."[18] In its bureaucratic language, the statement reveals what the debate was really about: protecting the status quo.

On February 11, 1988, the National Academy of Sciences rendered a verdict. It endorsed the idea of a genome project, which it estimated would cost $3 billion over fifteen years. But its approach was far more cautious than what the DOE had in mind. It recommended that the first order of business be making a map of landmarks across the genome that could herald the proximity of disease-causing genes. The ultimate goal—the systematic sequencing of the human genome—should be left for later, after scientists developed technologies to make this practical, the panel concluded.[19, 20] The panel noted that there was little precedent for spelling out very long stretches of human DNA. The largest continuous human DNA sequence that had been spelled out was the sixty-seven thousand base pairs of the human growth hormone gene. The academy team failed to foresee that the revolution in technology already powering the nascent personal computer industry might also be applied to DNA sequencing.

The academy report gave Wyngaarden the backing he needed to yank most of this large, politically visible endeavor away from the rival agency in which it was born. Wyngaarden promptly set up the Office of Human Genome Research (the name was later changed to the National Center for Human Genome Research) at the NIH to oversee the new genome effort.

To head the office, Wyngaarden needed someone who would be acceptable to the independent-minded biologists and also have the eminence to eclipse the DOE. The obvious choice was Watson, who was then the director of the Cold Spring Harbor Laboratory on Long Island.[21] Watson took charge in October 1988.

Officially, the NIH and DOE were coequals in this landmark endeavor.[22] But Watson's leadership along with the NIH's reputation as a biomedical powerhouse gave the NIH genome program visibility and prestige that the DOE couldn't match. Congress gave the NIH more money for its genome program—$28 million versus $19 million for the DOE in 1989.[23] And in 1990, the gap widened to $60 million for the NIH, versus $28 million for the DOE.

By the time Wyngaarden resigned as NIH director in the summer of 1989, the project had taken on a new hue. The idea of a massive sequencing project had cachet with the public, but now that the NIH had taken over, its scattered academic grantees were going to use the money for genetics research they felt was more important.

The DOE had planted this rose, but the NIH was planning to wear its blossom.

2

A REBEL LANDS A CAUSE

J. Craig Venter loves running into the waves. The bigger and more menacing they are, the more he likes to take them on. Perhaps he was born with an unusual temperament that can usher revolutions. As a child, he'd try to do what he pleased, delighting in battles against forces far too strong for him—like his parents. As a result, it seemed to him that he spent most of his childhood grounded.

While Venter's early environment may have contributed to his unruly behavior—being a second child and having "controlling" parents who had once been Marines—ordinary influences couldn't fully explain Craig's behavior. For reasons beyond even him, it seems, young Craig took rebellion to the extreme.

Venter was born in Salt Lake City on October 14, 1946, but spent most of his childhood in the San Francisco Bay area, where his father, John, worked as a certified public accountant and his mother, Elizabeth, raised Craig and his three siblings, two brothers and a sister. While Craig misbehaved, his older brother played the part of the perfect child, at least in Craig's mind, which was all the more reason for Craig to try not to be.

Venter likes to tell a story about how his high school English teacher was fired after being branded "un-American" because he slouched during the Pledge of Allegiance. As Venter tells it, his teacher's dismissal enraged him, and he led a massive sit-in that shut the school down for two days. He was called in to see the men's dean and asked why he was being so disruptive. "You must be getting a good grade from this teacher," the dean assumed.

"No. I'm getting an F," Venter replied.

"Then why do you like this teacher so much?" the official asked.

"Because I really *deserve* an F."

Craig was not a good student. He jokes that the only reason he received his high school diploma is that the school desperately wanted to be rid of him.

Having barely eked his way through high school, Venter didn't consider college. Instead, he meandered south to Newport Beach, where he worked at night as a clerk at Sears, Roebuck & Co. so he could surf during the day. But life was about to become more serious for Venter, just as it would for many other young men. By 1965, the Vietnam War had begun, and Venter was drafted into the army.

Venter had two weeks to report, so his father suggested he contact the navy to see if he could work out a better deal. Venter had been a competitive swimmer in high school and won a spot on the navy swim team, conveniently based in San Diego. This lasted only until the war heated up and President Lyndon B. Johnson dismantled all military sports teams. Venter had to figure out what he would do next. Having earned a top score on the intelligence tests he'd taken when he entered the navy, he had his pick of military training.

He decided to enroll in school to become a corpsman, the military version of a physician's assistant. Venter had an interest in medicine, but this was a risky move. Choosing to be a corpsman greatly increased a man's chances of getting sent to Vietnam.

In 1967, Venter was sent to a vast, makeshift navy hospital in Da Nang, a large port city in the northern part of South Vietnam. It served the region of Vietnam where the bloodiest battles of the war were fought. When he arrived, the hospital was strictly an all-male affair, run by corpsmen and doctors. Venter assumed control of the intensive care ward. He treated men whose legs had been blown off after stepping on mines, who had bullets lodged in their heads, or whose guts had been blown open. They were almost all Venter's age, or younger.

It wasn't the kind of work that lent itself to the clean, orderly hospital practices followed in the United States during peacetime. So when the navy nurses, who were all officers, suddenly arrived in fall 1967 and tried to enforce new rules to keep the place clean and organized, there was trouble.

Given Venter's recalcitrant demeanor, only magnified by the stresses of war, he wasn't just irritated by the petty concerns of the nurses. He was furious. "The [nurses'] biggest concern was cleaning the fingernails of POWs," he recalled. So one day, when a nurse objected to the way Venter, an enlisted man, was taking care of one of those POWs, he exploded. "Go fuck yourself!" he screamed at the officer—and ended up in the brig.

That's where thirty-year-old dermatologist Ronald Nadel found him when he arrived in Vietnam in October 1967. Venter was waiting to be sent into the jungle to treat soldiers in battle. If that happened, he was likely to be killed, because the Vietcong often targeted corpsmen to destroy morale.

But Nadel needed a corpsman for his dermatology clinic. He could immediately tell that Venter was the smartest kid in town. So Venter averted his sentence, and for the next nine months, he followed Nadel's orders faithfully, gratefully, and amazingly well. Nadel noticed that Venter rapidly learned to perform subtle tasks flawlessly. After looking at a particular skin condition in the microscope just a few times, the twenty-one-year-old corpsman would know what to look for on subsequent occasions. Nadel began to trust Venter's diagnoses as a matter of course.

Nadel told Venter he should go to college after the war, emphasizing that his intelligence was too great to waste. Venter was noncommittal, but he had been inspired by working with Nadel and by his visits to a Vietnamese orphanage where he treated children and their families for tropical diseases. He saw how minor medical assistance could change people's lives. Venter began to dream of making a career of traveling the world performing similar feats. He started talking to Nadel about medical school.

But before that, he had to finish his tour in Vietnam. And that wasn't all peering at skin infections under the microscope. In late January

1968, the North Vietnamese forces suddenly rose up throughout South Vietnam in what came to be known as the Tet Offensive. One night during the offensive, a helicopter hovered over the hospital firing tracer bullets down at Vietcong soldiers who were trying to force their way through the hospital grounds. Not expecting to survive the attack, Nadel sent a good-bye tape home to his family.

Meanwhile, helicopters landed in front of the trauma center, where Venter was working, delivering wounded soldiers. Others arrived by tank and truck—at the rate of thousands per day. For five sleepless days and nights, Venter sorted the soldiers who could still be saved from those who could not be, even as more bullets headed his way.

After Vietnam, life seemed more precious to Venter. Eager to make up for lost time, he enrolled in junior college and then raced through both college and graduate school, at the University of California, San Diego, in just six years instead of the usual nine or ten. By then, his dreams had shifted from medicine to science.

Venter tried to live each day as his last, as if, at any moment, a shell could fly his way and wipe him off the face of the Earth. He also began living without fear, at least without the ordinary fear that keeps many people from fulfilling their dreams. From watching the horrific demise of thousands of boys his age, Venter was acutely aware of the worst thing that could happen to him. Little else gave him pause.

In 1976, he moved to Buffalo, New York, for a job as a junior professor of biochemistry at the State University of New York. He studied a protein that serves as a receptor for the stress hormone adrenaline on brain cells and took several new graduate students under his wing. One of them was an exceedingly bright twenty-one-year-old woman named Claire Fraser, then engaged to be married to her college sweetheart, who worked in Toronto. Fraser thought her adviser was very distinguished. Venter was ten years her senior but barely out of graduate school himself.

Fraser stood out among her fellow graduate students not only because she was a tall, attractive brunette, but also because she was a star pupil. Driven and excited about her research, she began publishing papers and attending meetings very early in her graduate student years. Her fellow graduate students, mostly male, were struggling.

Envious, they treated her with scorn. Fraser had endured it all before. She'd had an interest in math and science in high school and college—and, in the early 1970s, females with such interests were pegged as oddballs. She was no exception.

But Craig Venter was not at all intimidated by Fraser's success. Instead, he was thrilled by it, and he became, as Fraser puts it, "my biggest cheerleader." Fraser was amazed, heartened—and strongly attracted. She hung on the professor's words: Don't be afraid of new challenges or risks. Embrace new techniques and technologies. Those are the seeds of progress. Years later, she said, "It's one of the best things I've ever learned." She was lucky enough to learn it through love, not war.

After Fraser finished graduate school in 1981, she and Venter married. Three years later, they both secured research jobs at the NIH's brain institute, the National Institute for Neurological Disorders and Stroke, or NINDS, in Bethesda, Maryland. It was a huge step up for both of them. The NIH was a prime place to do research. Budgets were many times fatter than they were at most university laboratories, and research was the focus, without distracting teaching obligations or committee work. The couple was happy to say farewell to the long Buffalo winters. Fraser was looking forward to gardening in a place where frost would never spoil her plants in early June.

But there would be frost of another kind at NIH, a kind that neither one of them predicted. Venter and Fraser continued to collaborate on their adrenaline receptor work. They were trying to work out a purification scheme that would enable them to pluck out a tiny amount of protein from a large mass of brain tissue, the first step in a long road toward finding the gene for the receptor. Ultimately, they wanted to explore how the receptor worked to transmit signals inside a cell. Such an understanding might help researchers develop better medications for high blood pressure, heart failure, and asthma through alterations in the signaling of adrenaline receptors in the heart or lungs.

In the 1980s, there were two main ways of finding genes. One way was to find a gene by examining the blood of members of a large family in which the disease was passed. Researchers would look for a fragment of DNA that was always inherited along with a disease. The gene for that disease was then probably on that fragment, thus narrowing the search. But Venter and Fraser's teams took the other approach. Instead of looking within DNA for the gene, they first looked for the protein molecule that a cell made from the gene. Proteins are strings of smaller molecules called amino acids, and there was a simple code by which one could translate a sequence of amino acids in a protein into a sequence of DNA bases in a gene.

In the early 1960s, two young NIH scientists deciphered that code. In the normal course of events, cells translate DNA into proteins through an intermediary called messenger RNA. The scientists made artificial RNA messages and plopped them in a solution that could turn them into proteins. By watching which proteins emerged from which RNAs, they figured out which RNA words specified which amino acids. In the code, sets of three RNA letters specified single amino acids. For example, AGC is the code for the amino acid serine.

But the code was redundant: there could be up to six triplet sets of RNA bases that code for a single amino acid. (Cells commonly choose from about twenty different amino acids while sewing the body's proteins.) Nevertheless, one can usually provide a reverse translation that is close enough to find the gene among the twenty-three pairs of chromosomes. Then the gene's sequence could be spelled out using the conventional methods of the day.

Venter and Fraser painstakingly isolated the receptor protein, identified its amino acid building blocks, and used the protein's amino acid string to determine roughly what DNA letters it must have come from. In 1986, Venter, Fraser, and their two laboratory teams together located the gene for the receptor, the first gene Venter found.

Finding the gene had taken them an entire decade, and sequencing the gene took a year by itself. Venter was frustrated by the slowness. There had to be another way.

There was, but at the time it wasn't clear whether it was a better way or just a much broader, bolder way. It amounted to spelling out, letter by chemical letter, the three-billion-base-pair treatise that was the human genome. From the letters, then, scientists would ultimately be able to sift out the genes, including those that underlie thousands of diseases plaguing mankind. Finding and understanding those genes might take a while, but knowing the human genetic blueprint would be the first step to deciphering it. In 1986, nobody was really doing this, but a few folks were dreaming about it. That dream was, of course, the Human Genome Project.

The project would rely on the breakthrough chemistry for sequencing DNA developed by Englishman Frederick Sanger a decade before. Sanger's invention took advantage of a special property of DNA's double-stranded structure: wherever an A existed on one strand, a T sat on the other, and vice versa. Similarly G's and C's were always matched. These hugging chemical bases kept the strands together. Understanding this property, Sanger grew DNA molecules in test tubes. He plopped in the essential ingredients: a DNA strand (the template), a special enzyme, and DNA's building blocks, A's, T's, C's, and G's. In the tubes, the DNA-growing enzyme stuck the appropriate bases to the template one at a time, laying down a T wherever an A existed on the template, a C wherever a G sat, and so on.

To determine the sequence of each strand, Sanger set up this process in four different test tubes. In each one, he tricked the enzyme by adding just a small amount of a stop-growing, or "terminator," form of one of the chemical letters, say, a T. The enzyme could add all four normal letters to the chain and the chain would keep growing, but when it happened to pick up one of the terminator T's, it would halt in its tracks. The result was a shorter chain that ended in T. The enzyme created lots of short strands of different lengths, all of which matched the template and ended in T's. In the other three test tubes, Sanger added terminator A's, C's, or G's. One of those tubes filled with shorter copies of the DNA piece all ending in A, another with copies ending in C, and the fourth with DNA pieces of varying length all ending in G.

In the last step, Sanger poured the contents of each test tube on a gelatinous sheet that separated them by size. In one column of the gel, the short strands that ended at the letter T were stacked one above the other, with the shortest strands at the bottom. In three other columns, the various strands ending in each of the other three chemical letters were stacked. With all the partial strands sorted by size and ending base, a person could, by stooping over the gel and squinting to see the tiny strands, determine the relative positions of all the T's, C's, A's, and

G's in the original template strand. That was the sequence of that piece of DNA.

Sanger's method was a breakthrough. For the first time, scientists had a way of deciphering the specific chemical code of life, heretofore hidden in a vanishingly small molecular world. But since humans had to read those gels with their ladders of stacked pieces, the process was very slow, tedious, and error-prone. Even by the mid-1980s, spelling out a human's entire genetic instructions in this manner was an almost unimaginable amount of effort, and experts predicted it would take at least three decades to complete the endeavor. No wonder many biologists thought this was a bad idea.

But Venter immediately took to it. It seemed the only way to get past this incredibly slow process of deciphering the molecular components of a cell. He liked the idea of tackling a big question rather than another minuscule one. As a man who now wanted every day of his life to count, Venter was getting frustrated by small questions. So as the Human Genome Project got off the ground, Venter was looking for a way to get involved.

He soon saw an entry point, a new technology that could change everything. It was an automated DNA sequencer—a machine that would relieve poor lab technicians of their job squinting at gels to decipher a DNA sequence. Caltech biotechnology wizard Leroy Hood unveiled the invention at the crux of the new device in 1986. It was Sanger's technique with a twist. Hood marked the chain-stopping base in each test tube with a fluorescent chemical. G's radiated fluorescent red. A's beamed fluorescent blue. T's burned fluorescent yellow, and C's glowed fluorescent green. All four test tubes could be spritzed onto a single column, or lane, on a gel. And as the fragments migrated to the bottom of the gel, a laser zapped the colors, which were read by a detector and fed directly to a computer. The computer translated the colors—say, red, red, blue, green, yellow—into chemical letters, G, G, A, C, T. Scientists still had to set up the reactions and the gels, but they didn't have to read the results.

A team at Applied Biosystems in Foster City, California, had built a DNA-sequencing device based on the fluorescent-dye approach—and Venter wanted one. Venter contacted the company to see what might be arranged. In 1987, with $110,000 in new grant money, Venter secured a prototype, dubbed the 370, and his lab became NIH's official test site for the device. Venter's lab team began testing in earnest.

Because his funding came from the brain institute, Venter focused on spelling out genes for proteins governing the communication between brain cells. These included genes for enzymes that manufactured chemical messengers called neurotransmitters and proteins that sat on nerve cells and received neurotransmitter signals. Interactions between such proteins lie at the root of our thoughts and emotions and form the basis for most brain medications.

The machines were still new, and Venter's team had considerable difficulty getting them to work reliably. Nevertheless, the group managed to generate a number of short bits of DNA sequence and assemble these into sequences as long as tens of thousands of base pairs, enough genetic real estate to span several genes. Within six months, the group had spelled out two new brain genes, a fourfold increase in speed over the old way. These were the first genes ever sequenced by an automated DNA sequencer.

Then, in 1989, the team began the first human sequencing test projects, focusing on regions of human chromosomes 4 and 19, those known to contain genes for neurological disorders such as Huntington's disease and myotonic dystrophy. At idle moments, Venter even dreamed of a futuristic factory that could eke out, chemical by chemical, the microscopic blueprint of a human being.

Soon Venter was ready to expand his enterprise to sequencing very large areas of DNA in hopes of discovering important new genes. Specifically, he wanted to decipher a big chunk of the X chromosome, one of the two sex chromosomes, stretching millions of base pairs and hiding a treasure trove of at least thirty disease genes. Since this was a much broader effort than NINDS typically funded, Venter asked the National Center for Human Genome Research to fund his audacious plans.

James Watson, the head of that center, was initially enthusiastic and, according to Venter, offered him $5 million one day in his office.[1] But because of mounting opposition to sequencing projects, Watson chose not to expedite Venter's request and instead required him to go through a more laborious process to get his money. In fall 1989, Venter submitted a formal proposal for his project that would be reviewed by a committee of his peers. Venter waited to see if the project would be funded.[2]

Venter was optimistic enough about the project's chances to interview a promising young University of Michigan graduate student for a post-doctoral position slated for X chromosome sequencing. In November 1989, Mark Adams flew to see Venter in Bethesda. A computer buff, Adams wanted a job that combined computers and biology. He found Venter's lab teeming with computers, not to mention friendly, creative, hardworking people. Adams was impressed, and so was Venter. Venter offered Adams a job on the spot without so much as checking a reference.

By then, Venter's laboratory was running like a well-oiled engine. By 1990, it housed four automated sequencing machines, more than any other lab in the world. The team was quite proud of their gadgets, bragging about them like teenage boys showing off the latest, loudest stereo equipment. And they were finally getting them to work reliably.

But despite the futuristic hum of Venter's machines and the energy that permeated his lab, the team was about to be shaken from its reverie. On Adams's first day on the job in May 1990, Venter got a call from Watson's genome center informing him that his lab wouldn't be funded to sequence the X chromosome after all. The anonymous referees who had reviewed his grant application thought the technology was too new to embark on such a massive effort. They were also concerned that Venter would not be able to link together a continuous string of millions of bases from short DNA strips, a problem that no lab had, in fact, yet solved.[3] It would not be the first time Venter would be slapped on the wrist for dreaming beyond his means.

3

GENE DARLING

At the time Venter was being rebuffed by the emerging genome establishment, another researcher three years his junior was anointed its newest darling. The thirty-nine-year-old Francis Collins, a fair-skinned geneticist who wore glasses and a thick mustache, had taken his share of professional risks. But the politically savvy Collins had found it far easier than Venter to work within the confines of the system. As he neared his fortieth birthday, Collins would receive one of the project's first major grants, and his star would only rise from there.

Collins was born on April 14, 1950, the youngest of four boys, to Margaret and Fletcher Collins. Collins's father had received a Ph.D. in medieval literature and eked out a comfortable life as a manager at Republic Aviation Corporation on Long Island. But before Francis was born, he and Margaret escaped to a farm in Virginia's Shenandoah Valley. So Francis grew up in an old log cabin surrounded by a hundred acres of pasture, fields, and woods that also sported a barn and outbuildings.

His intellectual father learned to milk cows, drive a team of white workhorses named Joe and Star, and raise sheep, the family's specialty. However, his hand-driven mode of farming was far from lucrative. So Fletch secured a teaching job at Mary Baldwin College in the nearby town of Staunton, and he joked that his teaching job was the real cash crop.

Francis was home-schooled until age ten. His mother would get up with him early in the morning to read stories and plays, invent games, plant vegetables, explore the hundred acres, find and count things, hike, or do whatever seemed sensible at the time. She used flashcards, dice, and number games to teach him arithmetic and eventually had him make detailed maps of the farm, using his math skills to draw them to scale.

Francis helped farm, too. He fed and watered the sheep and some-times carried newborn lambs from the barn to the fireplace in the house to warm up in winter. He also participated in a summer theater group that his parents had started in their property's grove of ninety-foot oak trees. At age five, he acted in *A Midsummer Night's Dream*. At seven, he wrote a 140-page script for *The Wizard of Oz* because he wanted to play Lion. Francis's father taught him to write music so he could write down the pieces he composed by ear on the family's pump organ. Francis became comfortable under the spotlight.

Like Venter, Collins was a gifted learner, but unlike Venter, he was thirsty for knowledge from an early age. Francis often said: "Show me how to do this," and "How does that work?" Perhaps because of this curiosity, he managed to learn enough from his haphazard early education to enter public school as a seventh-grader at age ten—on track for graduating from high school at age sixteen, which he did, as valedictorian.

Francis learned to balance various projects simultaneously from his father, who prepared for his own classes while milking his cows. In high school, Francis spent many late nights typing from the Dicta-phone in the office of a local pediatrician who had asked him to keep his records. Once the police barged in and questioned his presence there in the wee hours of the morning. In college, he toured with his parents' theater while handling a heavy course load.

At the University of Virginia, Charlottesville, Collins studied chem-istry, and, at age twenty-four, he earned a Ph.D. in physical chemistry from Yale University. But by then he had begun to find the subject, particularly the theoretical form he'd pursued, unsatisfying. It was too far removed from people, like those he had grown to love on his fam-ily's farm. One day, as he was finishing his dissertation, Collins phoned his parents and told them he was switching careers. He was going into medicine.

Collins had disliked biology in high school, in which teachers had their classes memorize the parts of a crayfish and other lists of names. He saw no logic or beauty in that. Nevertheless, he trudged off to medical school at the University of North Carolina, Chapel Hill, hoping at least to find satisfaction in treating people.

But he found more than that. It was December 1973. An austere pediatric geneticist named H. Neil Kirkman was giving the first-year students six lectures on genetics. He was not a flashy lecturer, but he presented principles to ponder, not details to memorize. The logical rules governing DNA code with its simple four-letter alphabet appealed to Collins's analytically trained mind. For the first time, life made sense.

Kirkman also brought patients with genetic diseases to class—vivid illustrations of how minute changes in a long string of DNA letters could produce complicated illnesses. One day, a man with sickle-cell anemia, a serious blood disorder resulting from a flip of a single DNA base pair, came to class. Another day, Kirkman brought a baby with a metabolic disease that led to mental retardation and liver damage—again due to a tiny genetic quirk. Collins was hooked. That was when he first thought that he might study these conditions and find answers to help those who suffered from them.

In Chapel Hill, where Collins stayed for his residency, he treated many adults with cystic fibrosis—a disorder inherited by one in two thousand Caucasian children—since the school had a long-standing interest in the disease. These patients produced abnormally thick mucus that clogged up their lungs and caused infections. The infections would destroy their lung tissue and eventually lead to death.

Despite years of research, no one understood what produced this mucus. One clue: The way the disease was passed between generations indicated that a single defective gene caused it. (In this case, the trait was recessive, meaning a child had to inherit defective copies of the gene from both parents to get the disease.) Many researchers were eager to find this gene. It would be the first clue to the mechanism of how the mucus was produced. As a resident, Collins found the clinical manifestations of cystic fibrosis fascinating, frustrating, and puzzling, but he had no plans to seriously study the disease. At the time, that seemed far too difficult. There seemed to be no way to attack the problem.

Even so, after his residency, Collins set off to earn his credentials in medical genetics at Yale in 1981. Luckily for him, it was just before the field was about to explode.

∞

AN ASTOUNDING DEVELOPMENT was about to rock the genetics community. It stemmed from the heroic efforts of a young psychologist named Nancy Wexler, who was on a crusade to unravel the mysteries of a horrific disorder called Huntington's disease. Huntington's is a degenerative disorder of the nervous system that produces not only wild, uncontrollable movements but also deranged thoughts and emotions— fitful anger, irritability, depression, and even dementia. It is invariably fatal. Wexler's crusade had been going on for more than a decade, ever since her mother had been diagnosed at age fifty-three.

At first, Nancy's mother would twitch a finger or a toe and occasionally stumble, but later, she displayed little control over her movements at all. It became extremely difficult for her to bring a spoon to her mouth, and the attempt often smudged food on her face. When she sat in a chair, her head often hung to one side and her legs would constantly change position. Her speech became unintelligible, and she became distraught over her inability to communicate.[1]

Like cystic fibrosis, a single gene caused Huntington's disease. But unlike CF, the Huntington's disease gene is a "dominant" gene, meaning that just one defective copy of it (of the two copies everyone carries) will produce the disease. A child could inherit Huntington's if *either* parent carried the trait. Nancy Wexler's chances were fifty-fifty.

Wexler hoped finding the gene would lead to a cure.

When the search for the gene began, it was conducted the old-fashioned way. Using an approach like Venter's, scientists were trying to get at the gene through the aberrant protein it made. They looked for a telltale molecular glitch in the cells collected from Huntington's patients—a key difference between the proteins in diseased versus normal tissues. The search was akin to finding a grasshopper in a vast field of grass. There was no systematic way of picking out the original assault—the primary cause of the problem—in cells plagued by copious signs of disease.

But in the 1970s, there was no alternative. Nobody had any way of finding where within the three billion human DNA bases lay an unknown gene stretching a few hundred to a few thousand bases. If the human genome were long enough to circle the globe, a gene would be shorter than a football field, and a disease-causing defect nearly as narrow as a grain of rice. Such a task was like finding something no bigger than a rice grain on the circumference of the globe.[2]

Nevertheless, finding an obscure genetic defect in the genome became almost tractable at the end of the decade, when researchers first envisioned a profitable way to locate unknown genes via nearby landmarks or "markers" on a chromosome. It is a process somewhat like giving directions to a friend's house by telling someone to watch out for a McDonald's a block away. We easily recognize McDonald's, and, when we do, we know we are close to our destination, even though we don't know what the friend's house looks like. Similarly, scientists can detect certain markers on a chromosome using biological techniques, but they have no way of detecting a disease gene directly.

The new gene-sleuthing method grew out of an idea that two young biologists, one from Stanford and the other from MIT, spontaneously advanced at a conference in Alta, Utah, in 1979. This snowy Utah town was also the site of the later conference that inspired DeLisi to start the Human Genome Project. In Alta, the two scientists—Ronald Davis and David Botstein—proposed a way of locating a large number of markers on chromosomes, like so many molecular McDonald's, any one of which might herald the proximity of a gene of interest.

One could detect all of these signposts by carefully manipulating a person's blood—sifting out the cells containing chromosomes, and then breaking up those chromosomes into pieces in a way required to reveal the markers. To find a signpost near a disease gene, researchers would test for lots of them in the blood of related people, some of whom have a disease, and look for a marker that was inherited consistently with the disease.[3] Such a marker is very likely to be close to a disease gene because of the way chromosomes are passed between generations.

In most of the body's cells, chromosomes come in pairs. But when eggs and sperm are made, they are endowed with just singleton chromosomes. Each chromosome in an egg or sperm cell is a mosaic of the pair from which it came. As an egg, say, is formed, each of the mother's chromosomes exchanges DNA with its pair in a kind of genetic do-si-do. Due to such "crossing over," or genetic recombination, parts of a parent's chromosome are often separated in the next generation.[4] This happens very frequently if those parts are far away from each other. However, neighboring segments of DNA are separated only rarely. They tend to travel together through the generations. Thus, if a DNA landmark is consistently passed on with a disease in a large family, researchers can assume it lies near the disease gene.

It was a fascinating idea, and it made sense. This was just the kind of logical, statistical thought process that drew the young Francis Collins into genetics in the first place. But at that time, the method proposed by Davis and Botstein was only theoretical. Nobody had ever located a disease gene this way, and many thought it was an incredible long shot. Top people in the field predicted that navigating by markers to find a gene with no other clues would take a decade; some said half a century!

But Wexler thought it sounded like her best chance of finding the gene for Huntington's. She arranged to collaborate with geneticist James Gusella at Harvard Medical School, whose lab would do the actual gene sleuthing, and then set off to the one place she knew she could find blood from which the secrets of Huntington's could be extracted: Venezuela.

In March 1981, Wexler traveled to the marshy borders of Lake Maracaibo, a world where women bore a dozen children each, families lived in houses on stilts, and men fished in long, narrow boats on waters that opened to the Caribbean Sea. Among the several thousand villagers in these lakeside towns, Wexler could make out some who staggered when they walked as if they were drunk. One man would suddenly fall and then pick himself up; another's gaze wandered erratically as her eyeballs rolled and flicked in their sockets.[5]

Like a plague that might have infested the water, Huntington's disease tormented these large Venezuelan families. Wexler's team needed families, since the chromosome landmarks could herald the presence of a gene only within a family. If one compared unrelated people, which chromosomal landmark traveled with the gene would vary, since the landmark had nothing to do with the disease itself, and that would confuse matters immensely.[6]

Wexler also needed large families. The families in the United States with Huntington's disease were too small to give researchers the statistical power they needed to be confident that a part of a chromosome linked with the disease actually contained the disease gene. But the Venezuelan clan was ideal. Wexler and her Venezuela team drew a detailed pedigree showing the vast network of familial ties. They drew blood from the family members and shipped it back to Boston.

At Massachusetts General Hospital, Gusella and his team extracted the DNA from the blood samples and then laboriously tested each sample for the presence and form of each marker. The researchers expected to have to test a hundred or a thousand markers before finding one that looked statistically promising as a beacon for a disease gene. But the Boston researchers had extraordinary luck. The twelfth marker they tested turned out to be linked to the disease.

The odds were better than a thousand to one that this particular marker was very close to the disease gene. Almost all of the Venezuelan family members afflicted with Huntington's disease had one form of this telltale marker, while their healthy relatives had another. Later, the researchers determined that the marker and, by association, the gene, lay on chromosome 4. It had taken just three years to narrow the gene's location to four million base pairs from three billion.

Francis Collins remembers the moment he saw Gusella announce his triumph at the 1983 meeting of the American Society of Human Genetics. "The world changed that evening in Norfolk, Virginia, when Jim stood up in the late-breaking session and showed his data," Collins later said. "All of us were stunned that it worked that quickly for such an incredibly important disease."

Researchers had taken a giant leap toward being able to find a disease gene without knowing ahead of time what sort of gene they were looking for. Many scientists, Collins included, jumped at this new opportunity to apply genetics to medicine.

❧

AT THIS POINT, nobody had yet found the precise genetic defect that caused Huntington's disease. They had merely located its neighborhood. Sifting through millions of base pairs to get to the specific gene was still considered by most people to be virtually impossible.

As a young fellow at Yale in the early 1980s, Collins recognized this problem and conceived of an innovative solution. He thought of a way of quickly homing in on a tiny genetic defect lying somewhere within a large chromosomal neighborhood. If he could get it to work, it could be a significant breakthrough.

At the time, researchers had to systematically scour the entire neighborhood for the gene, looking for particular bases that are different in people with a disease. Starting with one piece of DNA in the neighborhood, researchers would use this as "bait" to pull out an adjacent, overlapping piece from a pool of chopped-up chromosomal material. It's as if to walk along a sidewalk required building the sidewalk square by square. Before each step, one would have to find the next square of concrete by matching it up with the previous one.

As researchers slowly "walked" along the chromosomal sidewalk, they would test each piece of DNA for the presence of genes that might be disease related. The progress was painfully slow because the pieces were tiny, and one had to find each and every piece to make any progress at all.

Collins's trick: A way of "jumping" over a gap in the sidewalk to land on a piece of DNA some distance away. The technique was akin to searching for a name in the phone book by skipping from the A's to the D's rather than flipping one page at a time. It was a much faster way of crossing a chromosome to look for genes. A distance that took five years to walk could be jumped across in six months. But when Collins arrived at the University of Michigan in 1984, as an assistant professor, he didn't know whether the trick would actually work.

Collins inadvertently recruited his first graduate student by slamming a volleyball into his face at a genetics department picnic. Mitchell Drumm was slogging through his first of three required laboratory rotations and was thinking about where he would go next. He liked

Collins immediately and, on that basis alone, asked to join his lab, where he found himself assigned to make chromosome jumping work.

It took two years before Collins and Drumm mastered the technique. Luckily, by then, geneticist Lap-Chee Tsui at the Hospital for Sick Children in Toronto, Canada, and his colleagues had given them a disease to try it on.

A year before, in 1985, Tsui's team had mapped the general location of the gene for cystic fibrosis. Drumm had been thinking a lot about cystic fibrosis, since he'd learned that a former neighbor of his had given birth to a boy with this disease. So he and Collins decided to give chromosome jumping a test run in the region of chromosome 7 thought to contain the CF gene—a large region spanning some two million base pairs of DNA.

The group needed a starting point for walking and jumping, so they secured a piece of DNA in the CF region from researchers at the National Institutes of Health. They used that piece to pull out adjacent pieces—walking—and pieces farther along the chromosome—jumping. After a year of traveling along the chromosome, genetic evidence from other laboratories told them they had found two pieces of DNA on either side of the CF gene. But those markers were still 1.5 million base pairs apart. They had no idea where within that large region the CF gene lay.

To get any closer, Collins's team needed families afflicted with cystic fibrosis in which they could test each new piece of DNA. With those, they would be able to tell how close they were by how often their piece of DNA was inherited by the people who contracted the disease. A piece very close to the CF gene would rarely be separated from the gene via genetic recombination. A more distant piece would be more frequently separated from the gene. Knowing how often recombination took place, they could approximate just how far they were from the CF gene.

But Collins didn't have access to families. Other teams that did, such as Tsui's, appeared to be making faster headway. To put his eggs in another basket, Collins began testing his jumping technique on another target: the gene for Huntington's disease.

Though outwardly upbeat, Collins was secretly very anxious. His whole future was riding on the gamble that these strategies were going to work. Few believed they would anytime soon. Collins nervously pondered the possibility that he'd have nothing to show for his efforts by the time he was up for a tenured job at Michigan in about a year. Time was running out.

This was particularly hard on Collins, who set high standards for himself in his career. He impressed gene therapy pioneer James Wilson, who was Collins's closest professional colleague in Michigan, as the most ambitious person he'd ever met. "He's a deceiving type A," Drumm said of Collins. "He seems so relaxed but he's intense."

Collins hid his anxieties from most of his students, among whom he was extremely popular. A line of them would invariably form outside Collins's office whenever the door was open. They were drawn to his extraordinary knack for explaining things and deep respect for curiosity. He never made a student feel dumb and indeed treated most of their questions with a kind of reverence. As a joke, Drumm once gave Collins a tape dispenser in which he had inserted numbers as a way of keeping the waiting students in order.

Collins tried to lighten the mood in the lab by celebrating birthdays. He'd create homemade birthday cards on which he would draw a stick figure and pen a funny poem about the recipient.

But a darker reality struck by spring 1987. Another research group in London, led by Robert Williamson of St. Mary's Hospital, reported evidence in *Nature*, one of the world's two most prominent scientific journals, that they were extremely close to finding the cystic fibrosis gene. Rumors circulated that Williamson's group had probably found it.

That cleared the field. Only Collins's lab and Tsui's pressed on.[7] Something didn't add up in the *Nature* paper, Tsui recalled thinking. "We thought he maybe hadn't found all the pieces." Collins soon discovered that Williamson's candidate gene was not the culprit. But Williamson had to be extremely close and only stubbornness, combined with optimism about his team's approach, made Collins stick with it.

A few months later, the competition from Britain drove the U.S. and Canadian teams together. Tsui approached Collins at a meeting in

San Diego after hearing Collins give a talk about his CF work. Tsui told Collins: "You're jumping from the wrong places." Tsui offered to get Collins closer to the CF gene.

Tsui shipped Collins two pieces of DNA that his team had determined were very close to the gene. Patients with cystic fibrosis in the families they studied almost always carried them.

With each jump or step, the Michigan group pulled out a new piece of DNA and shipped it off to Toronto. Both teams searched every piece for signs of a gene. These included the presence of clusters of C's and G's, which often mark the start of genes, or a stretch of DNA letters shared by another animal, since gene sequences have remained relatively constant through evolution compared to the DNA between genes.

Early 1988 brought more good news: Williamson stood up at an international scientific meeting and announced he had, in fact, nabbed the wrong gene. He had been unable to find any mutations in that gene in patients with cystic fibrosis.

With renewed enthusiasm, Collins, Tsui, Drumm, and a Toronto biochemist named Jack Riordan continued their search for genes in the region. They found a few, but none of them was the one they were looking for.[8] At one point, they landed on Williamson's gene and moved on.

They knew they were close. They were running out of room on the chromosome. Back in Michigan, Drumm pulled out a piece of DNA after a jump they called E 4.3. The Toronto group had found it, too, and soon saw signs it contained a piece of a gene. Since the stakes were so high, the result was kept under wraps. Collins wouldn't even tell his closest colleagues what was going on. Wilson eventually quit asking Collins about the work.

The next steps in the project were done largely in Toronto, and suddenly even Collins's group was out of the loop. For several months, Tsui was tight-lipped about what was going on in his lab, though he once called Collins with a tease. They had "very interesting results," he said, but wouldn't specify what those results were.

Nobody in Michigan knew what to make of Tsui's vague utterance, and the tension mounted as the Toronto group worked on in secret. Some members of the Michigan team began to feel betrayed by the silence. They assumed Tsui was trying to take most of the credit for future discoveries.

Eventually, Collins figured out the reason. Administrators at the Toronto hospital were worried that Collins might leak critical information about the CF gene work to a small biotechnology company called Genelabs that had licensed Collins's chromosome-jumping technique. They thought there was some chance that Genelabs would find and patent the gene first. Tsui urged Collins to clarify his relationship with the company so that he could reopen their friendly dialogue.

Collins looked into the matter. He asked Genelabs whether they had any interest in cystic fibrosis. As Collins suspected, they did not, and Collins explained that in a long letter to Tsui. He pleaded with Tsui to reinstate their open collaboration.

Tsui agreed, and he told him what Riordan had discovered over the winter: the gene snippet in E 4.3 was used to make protein in sweat glands, a hint that it was part of the CF gene. The Michigan researchers began to get excited but dared not get their hopes too high.

More findings trickled in. In early 1989, the Toronto team discovered a mutation in E 4.3 that was present only in sweat gland cells from CF patients, not in healthy tissue. It was a deletion of three DNA letters, something that is almost never benign. Over the following months, the Toronto team found other parts of the gene and assembled those like a puzzle into one gigantic gene. It looked like the right one, but they couldn't be sure until they looked for the mutation in children with cystic fibrosis and their families.

One rainy night that May, while Tsui and Collins were attending a genetics meeting in New Haven, Connecticut, they hovered over a fax machine in Tsui's room in the Yale dorms, waiting for the latest data from Tsui's lab. As the paper fed through the machine, their excitement mounted. The three-letter deletion, the data showed, occurred in most of the chromosomes from patients with CF, but never on a normal human chromosome. The evidence was overwhelming.

Back in Michigan, Collins burst into Wilson's office and said: "I think we have it."

On August 24, 1989, the researchers held two separate press conferences, one in Toronto and one in Washington, to announce their landmark discovery. They flew between the two countries in private jets provided by the prestigious Howard Hughes Medical Institute, which supported Collins's research.

Reporters impatiently awaited the release of the scientific papers on the work, to be published in the September 8 issue of *Science*, to great fanfare. Though Collins's lab received less credit than the Canadians in the author lists of their papers—the Toronto team did well over half the work—Collins got more than his share of the publicity. His charisma and comfort in front of a crowd drew the public's attention. At least in the United States, people began to associate the outgoing, down-to-earth American with the discovery of the CF gene.

For patients with cystic fibrosis and their families, the gene's discovery seemed like a godsend. They had high hopes that the finding would lead to a cure for the disease. When Wilson walked into the annual meeting of the Cystic Fibrosis Foundation that October to talk about gene therapy for the disease, he felt the thrill. "It was electrifying," he recalled. "They were hanging on every word I said. It came close to a religious experience."

For Collins, however, it wasn't so much about cystic fibrosis as about a new science for getting a handle on genetic diseases, one he had termed "positional cloning" at a dinner in St. Louis after the paper's publication. "CF was the proof of the principle that you could in fact sift through three billion base pairs and find a very subtle change responsible for a rare disease," he later said. "It was truly exhilarating."

Positional cloning held promise for thousands of diseases. But it was totally impractical to do it the way Tsui and Collins had for CF. "So incredibly painful and laborious that was, with the rudimentary methods we had back then for looking at genome," Collins recalled. There had to be a better way. To Collins and like-minded colleagues, that way was taking shape as the Human Genome Project.

4

CROUCHING TIGR

Soon after James Watson took over as head of the NIH's new Center for Human Genome Research, he postponed the official start of the Human Genome Project to 1990, the first year the NIH would give out grants under the project's aegis. This pushed back the project's completion to 2005 from 2001, DeLisi's original goal. That seemed more realistic to Watson, who was worried about having enough time and money to complete the undertaking. Just over five million human DNA base pairs, or about one one-thousandth of the genome, had been sequenced in 1990, and there seemed little chance of deciphering the rest in less time.

The project's first five-year plan recommended avoiding large-scale sequencing entirely for the first few years. It did endorse small-scale efforts to sequence defined regions of chromosomes known to contain interesting genes. At Houston's Baylor College of Medicine, for example, researchers sequenced the part of the X chromosome containing a gene for Lesch-Nyhan syndrome, a disease of self-mutilation in boys. But the immediate focus was on developing methods that would lower the cost of sequencing to 50 cents per base pair, and on sequencing ten million bases of contiguous DNA, or 0.3 percent of the genome. A Manhattan Project for sequencing it wasn't.[1]

Large portions of the initial genome budget went to gene finding, a valuable effort but one that was already well funded by other NIH institutes. Twenty percent of NIH's impressive genome coffer of $108 million for fiscal year (FY) 1991 went to mapping landmarks on the chromosomes that could guide such gene hunts, versus just 6 percent for sequencing technology development and 1 percent for human sequencing itself. Forty percent was directed at a spectrum of efforts known as "physical mapping," in which researchers figure out how to break up the genome and put it back together again. Much of this money also was directed at disease gene hunting. So in 1990, Collins's laboratory got $1.6 million to fund its gene-finding work, just as Venter's request to do substantial sequencing of the X chromosome was rejected.

But Venter was not about to scale back his dreams. He was itching for something big and new. He flew off to Japan for a meeting wondering what he was going to do with Mark Adams, his brand-new postdoc. On the plane home, he got an idea. He had been thinking about the slow pace of sequencing efforts when suddenly he thought: Why wander through junk DNA when we could simply ask cells where the genes are?

Each cell in the body uses only certain portions of its DNA to produce proteins—the genes. And every cell advertises which snippets of DNA it actively uses by leaving transient molecular footprints—so-called messenger RNA—of those genes. Other researchers had already figured out how to isolate these fragile messenger molecules and, using special enzymes, make them into more stable DNA copies, so-called cDNA, for "copy" DNA.

By spelling the bases in cDNA molecules, Venter realized he could use cells to edit the genetic material for him. This would lead him directly to genes, enabling him to avoid sequencing any of the intervening "junk" DNA. And genes were the plum portions of the genome, the bits of DNA that were critical to the function of our bodies, the color of our hair, how high we can jump, or how vulnerable we are to colon cancer. Venter thought he could draw up the essential parts list for people, a basic starting point to understanding how bodies work and how to intervene when they fail to work properly. He wasn't going to make rough maps of the genome first or wait for better technology. Impatient as always, Venter thought he'd start right away.

After he landed, Venter ushered Adams and three of his other staff scientists into his office. He told them his inspiration. They thought it was a little nutty. The conventional wisdom was that cells produce RNAs in a biased way. That is, cells work a few genes like mad, using them to stamp out lots of RNA footprints, but they use most genes so sparingly that their footprints would be invisible. At least, it looked to be this way from studies of tissues such as skeletal muscle and blood in which only a very small number of different proteins seemed to compose almost all of the tissue. Adams and his coworkers figured they would end up sampling the same few boring, abundant genes over and over again, expending lots of effort to find only a tiny number of genes—perhaps less than 10 percent of the total.

For this reason, Venter's idea had long ago been conceived, considered, and rejected. British molecular biologist Sydney Brenner—the man who would then have relegated blind human DNA sequencing to felons—suggested sequencing cDNAs at Charles DeLisi's original genome project meeting in Santa Fe. But other scientists dismissed it as a waste of time.

But Adams was still game. It was an easy enough experiment. The laboratory already had several cDNA libraries—little vials of liquids containing cDNA from an assortment of brain cells, from which RNA footprints had been extracted and made back into DNA. They'd ordered these for experiments in which lab scientists were searching for specific genes on chromosome 4. Now, however, Adams was just going to pluck out a hundred cDNAs at random and prepare them for sequencing in Venter's small sequencing "center."

Neither Adams nor Venter knew exactly what the vials contained. The codes and cellular roles of these chemicals had not been catalogued. It was the genetic codes of these brain chemicals that Adams was to find out.

To speed the gene-finding process even more, Venter's sequencing team didn't spell out whole genes from beginning to end. Instead they sequenced just a small segment, which they called an "expressed sequence tag," or EST, of each one. These segments stretched just three hundred to five hundred bases, about the length that Venter's sequencers could handle at one time. They indicated the presence of a gene and also could be used as molecular bait to later extract the entire gene if that gene looked interesting. Alternatively, if enough tags had been randomly sampled from the same gene, they might be reassembled to re-create most of the whole gene.

Venter and Adams were careful to harvest DNA solely from brain tissues because all of the lab's funds came from the NIH's brain institute. It was a good place to start. In addition to the importance of the brain, brain cells made a larger variety of proteins (using genes as templates) than cells from any other organ. Venter was also using his brain project to indulge his real passion, sequencing itself. He was out to construct one of the largest sequencing laboratories in the world.

Venter's skinny, exuberant postdoc worked hard on the EST project, meticulously mixing drops in tiny plastic test tubes and feeding their four glorious sequencing machines with cDNA pieces. In the first several months, Adams and the rest of the team had sequenced bits of a hundred human brain genes and had just begun looking up their sequences using a new search program called Basic Local Alignment Search Tool (BLAST), which had just been created by the National Center for Biotechnology Information. BLAST enabled Adams, with help from staff scientists Tony Kerlavage and Mark Dubnick, to quickly match the unknown gene sequences to any similar known genes and thereby gain clues to their functions.

One day in August 1990, Adams was going down his list of one hundred sequences, checking each one with BLAST, when he came upon something surprising. The search program revealed that one of the human DNA sequences was almost identical to a part of a fruit fly gene that tells a cell in a developing fly whether to become a skin cell or a brain cell. The fact that the top match was a fruit fly gene meant that nobody had ever discovered a mammalian version of it. This is really cool, he thought. But there was more in store. Just hours later, another sequence on his list matched another key fly-development gene. It was the first time anyone had discovered a human version of either gene. Adams was ecstatic.

But Venter still wanted to find a way to pursue his other main ambition: large-scale chromosome sequencing. In October, Venter and Watson finally agreed on a plan to replace Venter's rejected X chromosome proposal. In line with the focus on human sequencing directed at specific genes, Venter agreed to submit yet another extensive application for funding to sequence regions surrounding disease genes on several different chromosomes.[2]

During the fall, members of his laboratory kept working on pilot human chromosome sequencing with ongoing funding from the brain institute. But Venter was disappointed with the results. Sorting the genes from the junk in the sequence was proving more difficult than he had thought without cDNAs to point the way.

At the same time, Venter's EST idea was working better than anyone had dreamed. The cells Adams had so far examined left a vast number of visible genetic footprints for him to scoop up and identify. Nature had been far kinder to the Venter clan than critics had predicted. As a result, the group was about to win a genetic jackpot unlike any seen before in science.

Sensing this in early 1991, Venter decided that what he really wanted was not genome center funds for sequencing chromosomes but funds for generating ESTs for all body tissues. He envisioned finding every one of the estimated one hundred thousand genes in the human body this way.

The prospect was electrifying to a man who had spent a painful decade finding a single adrenaline receptor gene. Suddenly, here was an incredible shortcut, a way of pinpointing the genetic roots of scores of diseases within months. Unraveling the genetic codes of cells wracked by Alzheimer's disease, for example, might pinpoint the genetic causes of Alzheimer's. Or he might take asthmatic lung cells, diseased heart tissue, or a cancerous part of a colon and spell out the genes expressed there to determine the molecular roots of those ailments. On a more basic level, he envisioned finding genes important in smell by sampling nasal tissue, in strength by examining muscle, or in intelligence by spelling out genes expressed in certain parts of the brain. This was the discovery of a lifetime; Venter expected the world to greet it with tremendous enthusiasm.

But when Venter gave a keynote speech at the Human Genome Organization meeting that year, he did not receive the expected accolades. Though some researchers found his lab's EST work quite exciting, others were upset that they were tagging large numbers of genes in such an unconventional way. This effort was so productive that it might jeopardize the funding of researchers studying single genes or genetic pathways in a more methodical, traditional fashion. It was the first hint of the battle that lay ahead.

When Venter got wind that his revised chromosome sequencing proposal would actually receive several million dollars in funding, he begged Watson to let him use some of it to sequence ESTs from various body tissues. With those in hand, his researchers would better be able to interpret the DNA sequence from a chromosome after they'd found it. But his request was denied. Though Watson may not have personally rejected the request, Watson did not, in fact, support the idea of sequencing only genes or parts of genes on a massive scale. He did not want to divert resources from the Human Genome Project's main goal, which was making maps of all the human chromosomes and sequencing all of their DNA. He felt that catalogues of genes might be useful, but that they should be compiled later, after the main mission of the project was completed.

Venter felt snubbed. He believed his request was rejected because the EST method held such promise, and it was a threat to the Human Genome Project. He figured that Watson was worried that Congress would cut funds for the project if Venter's group found all the genes quickly. On April 23, 1991, Venter wrote to Watson to turn down the imminent grant. If he couldn't use the money for what he believed was right for science, to hell with it. He wasn't going to let the gene-hunting elite push him around.

That's when, in Venter's view, he became "public enemy number one" at the genome center. Venter had just turned down a multimillion-dollar grant on principle. A rare gesture, indeed, and hardly a welcome one.

Venter probably did not help his cause by boasting to *Science* magazine in June that the EST approach was "a bargain by comparison to the genome project." It certainly was cheaper, at an estimated cost in the millions rather than the $3 billion allotted for the Human Genome

Project. But any suggestion that it might replace the Human Genome Project was absurd. Venter's project was cheaper and faster in large part because its goals were much narrower. For instance, the EST method was quite unlikely to capture all human genes, and it did not even attempt to decode the genes of any nonhuman creature, as the genome project did.

⚬

IF COLLEAGUES WERE annoyed by Venter's grandstanding about the scientific importance of ESTs, they became downright furious when he appeared to imply that this work might also have commercial value. Venter did not mean to imply anything of the sort, but the world of commerce came calling.

It was a patent attorney at the biotech firm Genentech who first recognized the potential commercial value of Venter's growing catalogue of gene fragments, some of which Genentech might conceivably use as fodder for drugs. Genentech had already used the gene for human growth hormone to manufacture a protein drug to treat some forms of dwarfism, for example.

The Genentech lawyer asked the director of the NIH's Office of Technology Transfer, Reid Adler, whether there were any plans to obtain patents on Venter's fragments. Patents on inventions were important to companies like Genentech because they enabled firms to pay for exclusive access to the invention, giving a firm an advantage over companies without such access. In this way, patents provide an incentive for people to put basic science discoveries like Venter's to practical use.

That was the first Adler had heard about Venter's work, but the letter from Genentech piqued his interest. So later, in spring 1991, when Adler ran into Venter outside his office at the NIH—Venter had been looking for another NIH lawyer when he introduced himself to Adler to ask for directions—Adler responded, "So you're the Craig Venter I've been hearing about!" Adler suggested that he and Venter get together sometime to talk about the intellectual-property issues surrounding Venter's team's discoveries.

Adler paid a visit to Venter's laboratory a few days later. He was amazed and impressed that Venter and Adams had actually discovered pieces of genes with such a seemingly crude technique. After thinking it over, Adler suggested to Venter and Adams that the NIH try to obtain patents on the fragments.

Venter and Adams had never considered trying to obtain patents on their work. The public sector genome community as a whole had largely ignored the commercial value of its discoveries up until that point. Biologists and others often dismissed the idea that life, or molecules that confer life, could be patented. But legally, some life-forms and molecules derived from them clearly could be, as long as the creature or molecule was in a purified or slightly altered state. In 1980, in *Diamond v. Chakrabarty*, the U.S. Supreme Court ruled that a modified living creature, a bacterium genetically engineered to break down crude oil, could be patented. And patent claims for bioengineered versions of human proteins such as insulin had also been upheld in court. This was the basis for the entire biotechnology industry.

Venter and Adams were not initially keen on trying to patent their gene fragments. They were concerned that patents would inhibit further studies of them. But Adler explained that patent law did not prohibit people from doing academic research on patented genes or publishing such work. It only prohibited using patented materials for commercial gain without a license.

On the other hand, if Venter and Adams's gene fragments slipped into the public domain without patents, Adler argued, the full-length genes from which they came might be deemed obvious and thus unpatentable. In that scenario, drug or biotech firms would be unable to obtain an exclusive license, making them unlikely to invest in treatments based on the genes. Adler felt that the safest course of action was for the NIH to try to obtain patents on the ESTs, which it could then license exclusively to drug companies.

Adler knew that the patentability of Venter and Adams's DNA bits, which were not whole genes with clearly defined functions, was uncertain. At the very least, however, applying for patents on them would help clarify whether any loosely defined DNA sequences, such as those to be uncovered in the general mapping of the human genome, were appropriate fodder for patents. Adler felt strongly that this issue needed to be addressed—and what better way than to use Venter and Adams's work as a test case?

Convinced, Venter and Adams went along with the patenting plan. It had to be carried out quickly. They were about to publish approximately three hundred of their first gene fragments in the journal *Science* a month hence, and foreign patent law required that patent applications be filed before publication. Adler called in patent attorneys at an outside law firm to prepare the application, which listed several hundred gene fragments and Venter and Adams as the inventors. It was filed June 20, 1991, just before Venter's paper went to press.

Watson and others in the genome center were vaguely aware of the application, but initially they paid little attention to it. Watson was busy with duties at Cold Spring Harbor Laboratories, where he was still serving as president, and caring for a son who had become seriously ill.

Then in July, at a congressional hearing on Capitol Hill that was held to update Senator Pete Domenici on the genome project's progress, Venter mentioned that the patents had been filed. Watson retorted that it was "sheer lunacy" to patent mere fragments of genes. He was concerned that if bits of DNA sequence with no clearly identified function could be patented, biomedical research might be stymied by time-consuming litigation. In addition, such patents might undercut the ability of later scientists, who identified an entire gene and its function, to obtain a patent if that gene happened to include a bit of already patented DNA. (He apparently did not share Adler's concern that letting the gene fragments slip into the public domain might also render the complete genes unpatentable, depending on how patent law was interpreted.)

Watson then said that "virtually any monkey" could run Venter's sequencing operation. "I am horrified," Watson told Congress.[3] Venter was hurt, but Watson's impromptu statement wasn't meant as a personal insult to Venter or to disparage his work, which Watson generally admired. It was meant to protest the idea that strings of meaningless chemical bases—which he likened to letters any monkey could randomly bang out on a typewriter—were appropriate fodder for patents. A patent, Watson thought, had to govern something useful.

But other prominent scientists did belittle Venter's work as menial robotics, not real science, an argument reflecting the biological community's lack of respect for technology and for large-scale sequencing as a scientific pursuit. The American Society of Human Genetics and the Human Genome Organization issued statements deploring the decision to seek EST patents.

The visceral reaction of the biology community against the patent applications baffled Adler at first. He did not realize that patenting these gene fragments threatened to upturn their world. Many of them had spent years, if not careers, studying a single gene, the protein it encoded, and what that protein did. In that world, a patent on a gene would be a small reward, ownership of a small sliver of genetic code, for a great deal of effort. But now it looked like Venter's relatively superficial approach could suddenly give Venter and Adams control over a huge amount of the genome with relatively little effort.

As of June 1991, Venter's team had sequenced 347 pieces of genes, and by mid-1992, they'd flown past the one thousand mark, boasting more than twenty-seven hundred bits of brain genes in their arsenal. Though these bits almost certainly did not all represent separate genes (some of them were undoubtedly parts of the same gene), Venter had staked out a huge swath of genetic territory, considering that fewer than two thousand human genes were known in 1991.[4]

It didn't seem fair that Venter and Adams could own so much genetic real estate, especially if other scientists later did most of the work to identify functions for any of Venter and Adams's gene snippets.

Patent law provided some support for these concerns. The basic criteria for issuing a patent—whether for a chemical or a television set—are novelty, nonobviousness, utility (commercial potential), and enablement: have the inventors provided sufficient information in their patent application to enable a person to parlay the invention into something useful? Many of Venter and Adams's gene fragments were novel, and their discovery of them would be judged obvious or nonobvious relative to the state of the field. But they did not necessarily satisfy the utility and enablement criteria.

Patent applications issued on single, complete genes had previously provided a wealth of laboratory data backing up the claimed function of the gene in the body, data that clearly indicated a plausible utility. That information could also be used to enable others to put the gene to good use—for instance, to produce a protein such as human growth hormone or human insulin that could be used as a therapy for dwarfism or diabetes. By contrast, Venter and Adams's gene fragments were accompanied by little more than speculation about what the functions of their corresponding genes might be. As a result, these new genetic "inventions" were much further removed from practical use. Patent lawyer Thomas Kiley framed the issue eloquently in the August 14, 1992, *Science*: "These patents cluster around the earliest imaginable observations on the long road toward practical benefit, while seeking to control what lies at the end of it."

Many researchers also worried that the sheer number of patents would concentrate too much power in the hands of one entity, in this case, the NIH. While some biotech industry scientists supported the NIH decision to seek patents on Venter and Adams's ESTs, others were concerned that the NIH might use its leverage in licensing intellectual property rights to control drug prices, a practice some congressmen favored. There were also widespread fears that patents would restrict the ability of scientists to freely use and distribute important biological data, thereby inhibiting scientific research, as well as general discomfort with the increasing commercialization of biological research. A symbol for all this angst, Venter began to attain lone ranger status among genome scientists at the NIH.

But Venter had a prominent supporter. The Bush administration's new NIH director, Bernadine Healy, appointed in April 1991, was impressed with Venter's work. She'd even asked him for advice about setting up a genome research program on the NIH campus in Bethesda, in addition to the genome research the NIH was supporting at outside universities. Healy also understood that almost none of the concerns about patenting Venter and Adams's fragments argued against *filing* the patents. On the contrary, they showed how much the biological community needed a test case to help resolve some of these intellectual-property issues—especially in light of the slew of genetic information about to be unveiled in the genome project.

Healy could have taken the easy way out and acquiesced to the scientific community by withdrawing the patent applications. But she, in fact, did the opposite. In February 1992, Healy agreed to add some two thousand sequences to the several hundred already in the 1991 patent application. Healy also wrote an article that appeared in August 1992 in the *New England Journal of Medicine* defending the NIH's recent patent filing as necessary to settle many outstanding issues.

"This policy preserves our options and gives us time to explore thoroughly a series of complex legal, economic, and scientific issues," she wrote. "We decided to file to hold our place. We filed so that we could publish immediately. We filed because no one knows to what degree function must be tied to structure for the patent requirements to be satisfied. In short, we filed to resolve the matter."

∞

AS MUCH AS the patenting hoopla repelled Venter's academic colleagues, it greatly attracted biotech investors. Venter's phone began ringing. A number of venture capitalists offered him lucrative financial packages to try to get him to leave the NIH and start a biotech firm based on his gene snippets, which they felt might lead to useful drugs.

Venter and his wife could have used the money. The busy working couple didn't have a cent to spare, not even to hire a housekeeper. They definitely needed one; Venter's big dogs shed hair in clumps, and Venter seemed completely unwilling to do household chores. Fraser once waited to see how long it would take before her husband would break out the vacuum cleaner. After seven weeks, she could not stand it any longer and decided they would hire a housekeeper whether they could afford it or not.

But Venter didn't take seriously most of the people throwing money at him. And most of the offers that he did take seriously he rejected. Venter wanted to be a scientist, not a businessman. At one point, he toyed with a $70 million offer from the biotech giant Amgen to set up an institute at one of Amgen's research labs. Venter really didn't want to work directly for a company, hemmed in by the demands of corporate bosses. He needed an even better deal.

Then in early 1992, in the heat of the patenting debate, Venter attended a panel discussion about the value of genes and ESTs at the

NIH. It was the kind of gathering that typically draws a small group of patent attorneys, but given the politics then surrounding the issue, the hall at the NIH was packed—as was the overflow hall—and the scene was teeming with TV cameras and journalists. Among the panelists, largely patent attorneys, was one venture capitalist. Alan Walton, who had been a professor of molecular biology at Case Western Reserve University in Cleveland, fell in love with the emerging biotechnology industry in the early 1980s. By 1991, Walton had defected to Wall Street, joining a venture capital firm called Oxford BioSciences dedicated to investing in cutting-edge technologies in the life sciences.

After Walton's talk, Venter introduced himself. Might they get together sometime to discuss some of Venter's scientific ideas? Over the following few weeks, Venter impressed Walton with his vision, and his means, for rapidly finding human genes. Walton considered the ideas revolutionary and, moreover, a major threat to a pioneering, but failing, genomics company over which he was then presiding. The company, called Genmap, was geared toward finding genes underlying the propensity for diseases such as breast cancer. Venter had found a more efficient way to locate genes.

Walton invited Venter to join Genmap, but he soon realized Venter's vision was too big to be realized within the confines of the doomed firm. So he and Venter batted around other ideas, within Venter's very narrow constraints. Venter told Walton he wanted to win a Nobel Prize, not form a company. But he wouldn't mind leaving the NIH and making a pile of money. Walton kept telling Venter that the two goals were not compatible.

Venter said he would seriously consider leaving the NIH to lead his own not-for-profit research institute, like the NIH but funded by venture capitalists. Walton was stumped at first. No venture capitalist in his right mind would consider funding a not-for-profit institute. It just didn't make sense financially.

Then he had a clever idea. What if they linked such an institute to a biotech company? The company would patent and license the institute's discoveries, providing a vehicle for profit. That might attract some funding. Venter could lead the institute and somebody else could lead the company, shielding Venter from the business and its threat to his scientific purity.

It sounded like a workable plan, but Venter required a lot of money to carry it out—more money than Walton had available at Oxford Bio-Sciences. Venter and his wife, whose lab would join his in the venture, had what were essentially permanent positions at the NIH with very generous research budgets. No matter how exciting the research was, neither of them would walk away from that for a small amount of money. They needed enough to support both of their research teams for many years—like the kind of money they'd been offered by Amgen. The typical amount of capital doled out by venture capital firms at the time would last for maybe two years. That wasn't enough incentive to make the jump to the private sector.

Walton had an inspiration. He knew someone in the industry who was likely to see Venter's vision and who also had sufficient capital to get Venter's idea off the ground. The man, whom Walton knew from the venture capital network, was the late Wallace Steinberg, then chairman of HealthCare Investment Corp. (later called HealthCare Ventures LLC) in New Jersey. One afternoon, Walton phoned the firm and spoke to one of Steinberg's partners, Jim Cavanaugh. Walton told Cavanaugh: "I'd like to introduce you to what I think is the biggest idea of the century. I'd like you to meet with Craig Venter."

Within a week, Cavanaugh and another partner, Hal Werner, traveled to Bethesda to rendezvous with Venter at a Hyatt hotel. After spending an afternoon with the NIH scientist, Cavanaugh and Werner became excited. Back in New Jersey, they talked about this potential new project at the firm's weekly technology meeting, and Steinberg was enthralled. He immediately grasped the implications of Venter's work.

Here was potential genetic fodder for a huge number of new drugs, he realized, information for which drug companies would pay dearly. The genes Venter was finding could serve as templates for therapeutic proteins—natural proteins that the body makes but that are deficient or absent in some diseases or that might be useful in excess. Diabetics, for example, lack the protein hormone insulin. Cloning the gene for human insulin led to the mass production of the protein, which became a lucrative and life-saving treatment for diabetes. Similarly, the natural protein erythropoietin (EPO), which stimulates the production of red blood cells, had become a $1 billion treatment for anemia after its gene was cloned.

Now, though, instead of starting with a particular protein—like human insulin or EPO—that was already known to be a key player in a disorder, researchers could start with a huge database of genes. They wouldn't know the precise function of most of the genes, but they'd have clues, and they'd use those clues to pick out promising ones and test them in cells and animals for medically useful effects. Venter's method had the potential to lead to thousands of protein drugs, some of which might indeed turn out to be blockbusters like EPO.

And that wasn't all. Many of Venter's genes would offer key clues to another, even more lucrative class of drugs. Most big pharmaceutical firms do not focus on protein medications, because those have to be injected, limiting their appeal. Instead, drug industry scientists try to develop drugs made of smaller molecules that can be swallowed in pills. To make such drugs, scientists typically start with a body protein with a known role in a disease and experiment with it to find a smaller molecule that interacts with it to restore its normal function. It is critical to have such "drug targets" as a starting point for drug development, and Venter's stash of new genes could provide a vast number of them. Drug companies might thus be eager to buy this mother lode of genetic information. Steinberg felt it was the opportunity of a lifetime.[5]

Steinberg and his partners decided to ante up $70 million (later upped to $85 million) for a combination enterprise. Venter would open a nonprofit institute named The Institute for Genomic Research (or TIGR—pronounced like the wild animal). TIGR would be tied to a for-profit entity called Human Genome Sciences (HGS), which would commercialize Venter's gene discoveries. Steinberg told Venter about this astounding offer one day in spring 1992 when he and his partners happened to run into him at one of the companies their venture firm had founded in Gaithersburg, Maryland. Venter was amazed. It was an unprecedented amount for anyone to invest in a new biotech company, amounting to one-third of Steinberg's venture fund!

Steinberg's firm wasn't going to hand over all $70 million up front. To get HGS started, HealthCare Investment put up $7 million, and Oxford BioSciences and one of HealthCare's limited partners, Rho Management, together came up with another $3 million. The remaining $60 million would be paid by HGS to TIGR on a quarterly basis over ten years. So if HGS went bankrupt, TIGR wouldn't get the money.

But the sound of a $70 million, ten-year commitment was enough to get the attention of Venter, Fraser, and everybody in their labs. In addition to the research funding, Venter would personally receive 10 percent of the equity in HGS; TIGR would receive another 11 percent of the stock. Coming from academia, Venter and Fraser were bowled over by the big money, the big stakes, and the meetings with venture capitalists and attorneys.

Venter resigned his NIH post on July 13, 1992. He and Fraser moved their groups to a building in Gaithersburg that served as the temporary headquarters of TIGR, which had been officially established seven days before. Husband and wife had left the battleground of the still-raging patenting drama, and, along with it, the safe haven that provided a guaranteed salary for life and a good research budget.

In August, the Patent and Trademark Office denied the patents on Venter and Adams's ESTs. The preliminary decision meant little. Virtually all patent applications are initially contested, as patent examiners consider that part of their job. So the NIH appealed the decision and, in September, filed a new application on another forty-five hundred gene sequences. But the controversy surrounding these particular applications would be moot. After Nobel laureate Harold Varmus took over as the NIH director the following year, he withdrew them. He did not believe that patenting at this stage promoted technology development, and he thought it could impede important research collaborations. Every bit the basic scientist, he saw the sequences as research tools rather than commercial products. And so the question of whether such fragments were patentable was left for companies, like the newly formed HGS, to argue.

HGS had been formally founded on June 26, 1992, headed temporarily by Lewis Shuster, a businessman recruited from a diagnostic laboratory. From the start, its investors were very nervous because of the huge up-front commitment Steinberg had made to the firm. Steinberg's peers initially ridiculed his decision as "Steinberg's folly." At this stage, it wasn't at all clear how much commercial interest there would be in Venter's gene fragments. In the press, many scientists claimed they were useless. Colleagues worried that Steinberg may have just flushed $70 million down the toilet.

So Steinberg and Shuster very quickly began looking for a big drug company that would buy Venter's gene data. Steinberg, along with advisers to his firm, made elaborate presentations to pharmaceutical giants such as Rhône-Poulenc Rorer and SmithKline Beecham about the value of Venter's gene fragments for finding lucrative new drugs. One of Steinberg's advisers at the time was a prominent Harvard virologist, William Haseltine, whom Steinberg was hoping would take the job of permanent chairman and CEO of Human Genome Sciences.

Steinberg didn't tell Craig Venter this, knowing that Venter would want to pick the CEO himself. But Steinberg mentioned Haseltine, a man with considerable experience in biotech, as a potential consultant to HGS, which Venter was largely running at the time. On this pretext, Steinberg arranged for the two men to meet.

5

DRAGON SLAYER

Like J. Craig Venter, William Haseltine is strong-willed and ambitious. He is also the second of four children and once had a rebellious streak. Both were born with an unusual late-blooming intelligence that would prove to be sharp and powerful.

But these are about the only personality traits the two have in common. While Venter is casual and chatty, Haseltine is formal. A tall, imposing figure who favors dark conservative suits that blend with his neat, black, thinning hair, Haseltine is less likely to chat with acquaintances than to deliver a speech, embellished with evocative metaphors. He prepares answers in advance. It's part of his ongoing effort to complete a "hypothetical interview" that he hopes might ultimately encompass every question he might ever be asked. While Venter is glossing over details, Haseltine attends to a story's fine points and may insist on his interpretation of them.

People who meet Haseltine for the first time are often struck by how vehemently he trumpets his accomplishments, as if it were terribly important that the listener understand their magnitude without delay. Listeners figure that anyone this boastful has to be full of hot air. But there's a surprising amount of truth to Haseltine's claims, making his need to hammer them home even more curious. But perhaps that is just his way of trying to get more of the attention he seemed to have lacked as a child.

Haseltine is organized, efficient, and dedicated. But he has an expansive imagination as well. His view of the universe is a glamorous one. In it, nothing is out of place, and yet nothing is at all ordinary. He envisions a future in which illness is dead, killed by chemicals he helped craft.

To inspire his creativity, Haseltine spends a lot of time studying art. He visits museums and art galleries frequently and decorates his office and regal New York City condominium with paintings and sculptures from many different cultures and eras. In one of his favorite paintings, a five-hundred-year-old masterpiece by the great Renaissance artist Raphael, disease rears its head as a black dragon with wings clawing at the feet of a soldier astride a white horse. The warrior, a young St. George, clenches a long sword, ready to dispatch the monster.

Haseltine has long been looking for his sword.

☙

WILLIAM ALAN HASELTINE was born on October 17, 1944, in St. Louis, Missouri, where his father was stationed in the military until the end of World War II. But Billy, as he was called as a child, spent most of his childhood in the small town of China Lake, California, in the middle of the Mojave Desert.

Haseltine's most vivid childhood memories surround the battles against the dragon. One day when he was seven years old, a doctor detected an unusual grating sound emanating from Billy's heart and diagnosed him with a potentially life-threatening heart condition called pericarditis. The thin cellophane-like membrane surrounding his small heart was inflamed, squeezing his heart and restricting its action. The only treatment was bed rest. Billy lost months to his heart trouble. He eventually recovered, but the dragon had left its mark.

It left deep scars on Bill's mother, Jean, too. Before Bill celebrated his eighth birthday, his mother was hospitalized for a detached retina. In those days, reattaching the retina required draconian surgery in which ocular muscles were cut and parts of the retina were cauterized. Jean then spent weeks in bed. For a year, she had to wear pinhole glasses so she wouldn't shift her eyes to look around, and she was terrified of bumping her head for fear the retina would detach again. Indeed, it did do so once.

Jean also had a skin condition that caused her hands to blister and bleed. Sometimes the sores would become infected, and Bill could see the infection track up her arm: thin red lines extending upward like the mercury on a thermometer. The children were told that if the red line reached the chest, she would die.

One summer, when Bill was just nine, his mother was diagnosed with manic depression. Her condition was severe, and she twice committed herself to the hospital. When she was away from home, Bill's older sister, Florence, then a teenager, took charge of most of the household chores.[1]

Bill's father, William, was a physicist who worked on the ballistics systems for the Polaris and Sidewinder missiles at the U.S. Naval Ordinance Testing Station in China Lake. He worked hard at his difficult job and was very proud of his children. But when he got home after work, he preferred to be left alone. When the children disturbed his peace and quiet on weekends, Bill's father sometimes ran after them with one of his straps. If he caught up to a child, he or she would get a quick lash on the back of the legs. As a defense, Florence recalls that she and Bill would hide almost all of their father's straps, leaving out only the softest one.[2]

Bill was a late bloomer academically. Not until high school did he begin to excel. In 1962, Bill ventured over the mountains to enter the University of California at Berkeley, intending to become a doctor. Haseltine was a very serious student. He polished his shoes every evening before he went to bed, and in the morning, he dressed in a jacket and tie before going off to class at eight A.M. every day. He was determined to secure every advantage in the contest of college and life.

His first day of chemistry class at Berkeley, the professor told an auditorium full of students: "Look to your right. Now, look to your left. At the end of this year, only one of you three will be here." Bill was determined to be that one.

As it happened, Bill liked chemistry, and he frequently raised his hand to ask questions. He caught the eye of the chairman of the chemistry department, George Pimentel, who selected him and fourteen other freshmen from a class of three thousand first-year chemistry students to participate in a summer science program, where the students met a different Nobel laureate every week.

The following summer, Haseltine was invited back to Pimentel's summer science program, this time to consider the question: What is life? Evidence had just surfaced suggesting that life existed on Mars. It reported signs of chemicals in the Martian atmosphere called aldehydes, which had to be made by a living thing. Pimentel, who was then designing instruments for the first U.S. Mars flyby to be launched that November (1964), told Haseltine to try to reproduce the results using model atmospheres of Mars.

Instead, Haseltine questioned the scientist's results. He conducted some simple experiments and found an alternative explanation that cast serious doubt on the evidence for life on Mars. Pimentel helped Haseltine publish his work in one of the most prestigious scientific journals: *Science*. Haseltine was naïve about the accomplishment. He didn't realize that publishing in *Science* was a big deal.

Haseltine graduated among the top five in his class of eight thousand, and he was accepted to both Harvard's graduate school and its medical school. Torn about which to attend, Haseltine consulted two prominent Harvard neuroscientists about his quandary. Their advice: Unless Haseltine felt a strong need to treat people with his own hands, he would make a greater contribution by focusing on science. He took their advice and enrolled in Harvard's biophysics department.

He went to work in the laboratory of Walter Gilbert, the physicist-turned-molecular-biologist who would later win a Nobel Prize. James Watson helped Gilbert run the lab, a group of fifteen to twenty highly competitive and energetic graduate students, postdoctoral fellows, and junior and senior scientists.

In that intense milieu, Haseltine feasted his intellect on the mystery of how genes regulate the construction of complex creatures like people with their many types of cells. Genes help determine how legs, livers, hearts, and hair are made and arranged in the body—but not because cells from different body parts have different genes. With rare exceptions, all cells in a single body contain the same DNA. But some genes are turned off in each of our cells, and it is the genes that remain "on" and making protein that distinguish one cell type from another.

In the late 1960s, many scientists wanted to know how a cell turned on or off its genes in hopes of shedding light on how human cells functioned. Many researchers in the field, including those running the Harvard lab in which Haseltine worked, started by studying that process in bacteria, which are single cells.

Haseltine embarked on a thesis project involving one of these bacteria. He hoped to purify a protein molecule that was thought to make a gut-dwelling bacterium stop foraging for food and start multiplying like mad, a transformation that involved turning on new genes. But after months of effort, Haseltine saw his project evaporate as his experiments cast serious doubt about whether the protein in question was necessary for the transformation after all.

Frustrated, Haseltine tried to make a difference in other ways. He exposed the wrongs of the Vietnam War in articles for *The New Republic* and for a local newspaper called *Boston After Dark*. Among other things, his reports revealed the perils of "agent orange" and the bulldozing of large tracts of land in South Vietnam, which improved visibility but greatly damaged the local ecology.

Haseltine was also searching desperately for a new thesis. Soon he found one involving the same bacterium as before. This time, he was going to artificially produce and characterize a substance nicknamed "magic spot." This substance helped switch the bacterium from multiplying back to foraging, the reverse of the mechanism he'd pursued before.

For the next year, Haseltine almost never left the lab, except on Sunday afternoons to sleep. One night at about eleven o'clock, Haseltine went into a dark room to view the photographic results of his latest experiment. He pulled out the film and saw that he'd done it. There was magic spot! He knew then that he had a promising thesis project, since discovering what the substance was and exactly how it was produced seemed quite possible now.

But finishing that work was far more stressful than Haseltine had anticipated. A rival Harvard professor wanted to figure out how magic spot was made before anyone in Gilbert's lab did, and he assigned one of his graduate students to compete with Haseltine. The competitor even took a peek at Haseltine's lab notebook in search of helpful clues. When Haseltine complained about the unwanted adversary to his adviser, Gilbert shrugged. Competition was good for science, he said.

It wasn't good for Haseltine. The once-burned young man worked around the clock for nine months until the effort literally made him sick. He ended up in the hospital for a week, although it's unclear exactly what ailed him, aside from utter exhaustion. When he recovered, he made a sprint to the finish, this one lasting four months. Haseltine pieced together the precise chemical steps that led to the production of magic spot. He unraveled the mechanism that flipped a bacterium between its two lifestyles ahead of his competitors, and he received his Ph.D. by graduation day, 1972.

Haseltine's mentors, including Gilbert and Watson, taught him, he recalled, "not to be afraid to achieve what you imagine." Big questions would not frighten him. To Haseltine, the big questions were those that would make an impact on human health. And so he was eager to part ways with bacteria and study the cells of mammals such as humans. But in 1973, the tools for doing so were extremely limited. Working with animal cells was very difficult, as researchers were just learning to grow them in beakers and test tubes.

At that time, the best tools for studying animal cells were, ironically, viruses. Viruses are known to most people as a class of germs that can cause everything from nausea and chills to cancer and brain damage. But to scientists, they are tools because they have marvelously sinister and clever ways of sneaking into our cells and taking over their machinery—including, it turns out, their DNA. Haseltine astutely saw viruses as his key for getting into animal cells and learning how they function and malfunction.

With that in mind, he looked for "the smartest person I could find" to teach him to use those microscopic instruments. He chose thirty-two-year-old David Baltimore, then a biology professor in Cambridge at the Massachusetts Institute of Technology (MIT). Baltimore had already published the groundbreaking discovery that would eventually earn him a Nobel Prize. He'd discovered a strange protein used by some viruses to spin their own genetic material, made of RNA, into DNA.

Baltimore's protein was called reverse transcriptase, since spinning RNA into DNA is the reverse of what normally happens in cells, where DNA is used to make RNA.

For two years, as a postdoctoral fellow at MIT, Haseltine learned about the breed of viruses that used this unusual protein. They were called retroviruses. Some retroviruses were known to cause cancer by hijacking a cell's genetic machinery and making it copy itself like mad to create a tumor. The researchers could actually watch cells in lab dishes turn cancerous. As Baltimore became more interested in using viruses to get a handle on the origins of cancer, he decided to move to MIT's Center for Cancer Research, taking Haseltine with him.

Haseltine made major inroads in understanding how retroviruses hijack a cell's DNA, indelibly revising the blueprint for healthy operations. Unraveling such mysteries was especially fun for Haseltine when his findings contradicted those of others and he could show the other findings were wrong, just as he'd cast doubt upon evidence for life on Mars as a college sophomore.

Some of the data he gathered at MIT clashed with that of another young virologist named Harold Varmus, then at the University of California at San Francisco. Varmus would go on to earn a Nobel Prize and, later, to head the most powerful funding agency for biomedical research, the National Institutes of Health. But in the early 1970s, Haseltine found Varmus to be in error about one aspect of the viral hijacking process. "Harold said it was one thing, and he just got it wrong," Haseltine pointed out without prompting, a quarter century after the fact.

In 1975, at age thirty, Haseltine parted ways with Baltimore to set up his own laboratory at Harvard's Dana-Farber Cancer Institute, where he'd secured a post as an assistant professor. There, he and John Coffin, a young virologist at Tufts University School of Medicine in Boston, began a long-running collaboration to reveal how certain retroviruses caused cancer in mice. Their goal was to shed light on how human cancer starts. Though the collaboration was productive, it was not always easy for Coffin, as Haseltine's aggressiveness often clashed with the Tufts virologist's milder demeanor.

It wasn't long before Haseltine's outspokenness got the better of him. Not everyone agreed with the scientific conclusions Haseltine and Coffin had drawn in their virus work. When one of their biggest critics, virologist Peter Duesberg, visited Dana-Farber to judge the research as part of a grant review, Haseltine got into a screaming match with him.

Sure enough, the team's funding was reduced. The more senior professors on the team were furious at Haseltine. But it was a point of pride for Haseltine. "I was right and incontrovertibly right from the evidence," he later explained.[3]

Haseltine's tenacity also found more productive channels. At Dana-Farber, he arose early to attend the six-thirty to eight A.M. medical rounds, in which doctors and trainees discussed their patients, many of them children with cancer. This was Haseltine's chance to understand the physician's world, the world in which he planned to apply his scientific advances. Haseltine used that knowledge to help create a new department called Biochemical Pharmacology, geared toward applying advances in molecular biology to the treatment of human cancers. He also helped set up a program to train doctors to apply the techniques.

December 1981 brought another ugly face of the dragon—this one in disguise. The disguise was a new form of cancer, Kaposi's sarcoma, that had reared its head among gay men. Haseltine soon learned that this strange form of cancer might be linked to a disease of the immune system that was then spreading among certain segments of the populace. Acquired immunodeficiency syndrome, or AIDS, was then devastating its victims' immune systems, leaving them prey to all sorts of infections and cancers.

An expert in both cancer and diseases caused by retroviruses, Haseltine began to notice some interesting features of this strange new disease—features that might lead to clues to its cause. Haseltine had by then moved from mouse viruses to studying the first known human cancer-causing retrovirus: HTLV-1, for human T-cell leukemia virus, which Robert Gallo, then at the National Cancer Institute, and his colleagues unveiled in 1980. Haseltine noticed that like HTLV, the agent causing AIDS was sexually transmitted and infected immune cells called T-cells. It appeared to be a retrovirus.

Once the virus causing the disease was isolated, Haseltine's team set out to decipher its molecular workings so that drugs could be developed to gum them up. By late 1985, Haseltine had printed out HIV's entire genetic blueprint, chemical by chemical, on a sixty-five-foot strip of computer paper, which he and his graduate student Roberto Patarca wrapped around a large room of Gallo's lab in Maryland. It was ten thousand chemical letters long.

From that viral code, Haseltine's team plucked out many of the genes and figured out what they did. Gallo remembers Haseltine as "brilliant," and a tremendous asset to their collaboration. "We went faster because of Bill Haseltine, and that's what I call a good scientist," he later said.

Haseltine never forgot his childhood ambition to cure patients. From the beginning of his career, he was focused on the practical applications of his work. Starting in the late 1970s, Haseltine watched many of his academic friends, including his mentor Wally Gilbert, form or join biotech companies. He thought, If they can do it, so can I. He bought books on venture capital. He bought himself a suit and a briefcase. And he found himself an investment banker to help draw up a business plan for the firm, which he named Almar Scientific after his children, Alexander and Mara. This changed to Cambridge Bio-Science (and much later to Aquila Biopharmaceuticals). It would create animal vaccines. He recruited senior Harvard professors to help and serve on his scientific advisory board. In 1981, the company was born.

Its stellar scientific team created the first effective retroviral vaccine, which prevents cats from getting leukemia. Haseltine initially served as chairman of the company's board of directors and of its scientific advisory board, but he hired a senior management team to run the firm on a daily basis. Haseltine did not get along with the company's CEO, however. In the mid-1980s, Haseltine became disgusted by a strategic decision the CEO made that, in Haseltine's view, squelched the company's chances of becoming a major player in AIDS diagnosis. At that point, Haseltine resigned from the board and sold all his stock in the firm. The stock was then near its peak price, and Haseltine made millions of dollars in the transaction, giving the forty-one-year-old professor a taste of how lucrative biotech could be.

In 1985, a mutual friend introduced Haseltine to the man who would bring him back to business. Wallace Steinberg, who'd recently been the head of R&D at Johnson & Johnson, was then forming his own venture capital firm called HealthCare Investment Corp. Within two years, Haseltine was sitting on HealthCare's scientific advisory board.

Through the monthly board meetings, Haseltine and Steinberg became fast friends. By 1990 they were meeting in New York City every other weekend for lunch and a trip to a museum or art gallery. They'd talk about ideas for new businesses, and their conversations spawned several new biotech firms based on Haseltine's scientific ideas.

The first was the Virus Research Institute (now Avant Immunotherapeutics), established in 1990 to create new vaccines. That was followed by LeukoSite and ProScript, both aimed at developing anti-inflammatory drugs, and Activated Cell Therapy (now Dendreon Corp.), where researchers developed a way of harnessing the body's immune system to attack cancers. Each time, Haseltine received 5 or 10 percent of the initial capital of the company for conceiving it and helping set it up.

Then one Saturday early in the summer of 1992, Steinberg mentioned a new deal he had in the works. He told Haseltine it was probably the biggest idea for a company that he'd ever encountered. It might even get Haseltine out of Harvard, he said.

A biologist named J. Craig Venter at National Institutes of Health, Steinberg said, had developed a vastly more efficient way of finding human genes. Instead of creating ever more detailed maps of the entire human genetic blueprint—the plan of the government's Human Genome Project—Venter plucked out just the small genetic footprints cells read to make proteins. This edited genetic text comprises the most medically useful 3 percent of the genetic code, because it is this text that governs the workings of the human body.

And knowledge of those genes, Steinberg said, would give big drug companies the clues they needed to devise new breakthrough therapies. Since Venter recoiled at the thought of heading a company, Steinberg explained, they had settled on a deal in which Venter would head a nonprofit institute (TIGR) that would be tied to the for-profit HGS. TIGR would receive an unrestricted grant to discover human and other genes, and HGS would package Venter's team's genetic data and sell it.

Would Haseltine consider becoming the chief executive officer of HGS? Steinberg wanted to know.

Haseltine was intrigued. He knew something about Venter's work from reading his scientific papers. Haseltine also had been looking for a way to get involved in genomics, the study of genes and genetic codes on a large scale, a field he had earmarked as the wave of the future. Here, perhaps, was his chance.

6

THE ODD COUPLE

William Haseltine traveled from Boston to Bethesda to meet Craig Venter for lunch in July 1992 at the Bethesda Plaza, a cafeteria in a mall near Venter's office at the NIH. As the two men sat outside in the sun eating hamburgers and Cokes on plastic trays, the balding, casual Venter produced a few computer printouts showing his companion how critical information could be gleaned from a mere fragment of a gene. Haseltine immediately saw the implications.

What were Venter's plans for HGS? Haseltine wanted to know. To Venter, who developed the idea with Alan Walton, the firm would sell big drug companies access to the huge data bank of genetic information that TIGR would create. This data would give its customers a head start on developing new drugs. HGS would not make drugs itself. That was too risky and costly; it takes hundreds of millions of dollars to bring a drug to market, and only a small fraction of experimental drugs make it. Venter initially envisioned HGS as a technology transfer office for TIGR that would broadly license TIGR's data.

Haseltine listened politely, but he had his own ideas. Haseltine thought the Human Genome Project leaders were largely neglecting the medical applications of the new DNA technology, and he wanted to create what he later called "the first and most successful spin-off of the Human Genome Project." On the plane back to Boston, Haseltine scribbled his goals on the back of an envelope.

The Goals and Objectives for the New Company

Goal: To build a global biopharmaceutical company that discovers, develops, manufactures and sells its own protein and antibody drugs

To achieve this goal there are four strategic objectives
1. To be first to discover the great majority of the human genes in their useful cDNA form
2. To create systematic methods for turning knowledge of new genes into proprietary knowledge of their medical use
3. To obtain near and long-term funding by creating partnerships with large pharmaceutical companies
4. To build the infrastructure required to discover, manufacture, develop and sell our own protein and antibody drugs

Reporting back from his lunch, Haseltine told Steinberg he wasn't interested in heading a licensing firm, but he wanted to build a world-class drug company. He knew that fewer than one in a hundred new ideas for drugs lead to a marketed medication, but he'd compensate for that risk with a virtually limitless supply of ideas. He was confident that among the thousands of proteins circulating in the human bloodstream, there were dozens of useful drugs like insulin and erythropoietin. Getting just one or two of them to market, he reasoned, could financially sustain further drug development.

He told Steinberg: "If you want me to do this, I'll consider it." But there were conditions. Ever since his conflicts with the CEO of his first firm, Haseltine had vowed that his would be the deciding voice on key strategic decisions in any future company. He told Steinberg: "Everything I tell you to do, you do, or I quit." Haseltine also wanted to build his own scientific team to make up for what he saw as Venter's weakness. He regarded Venter's discovery as extremely useful and even "the key to the future of medicine," and he was impressed by Venter's energy and brilliant promotional skills. But Haseltine considered Venter a mediocre scientist, one with few insights other than how to sequence DNA.

Haseltine breezily contemplated whether he needed Venter at all. He knew that Steinberg, an honorable man, would never think of cutting Venter out of the business at this point. In addition, Venter would be useful for getting their DNA-sequencing facility running quickly. But, he confided years later, "I knew we wouldn't need him for very long." To Haseltine, Venter's institute was like a booster on a rocket, an engine that would propel him for a while but then fall away.[1]

Venter had taken an instant dislike to Haseltine. He found him slick. Haseltine vaguely reminded Venter of Gordon Gecko, the high-powered speculator played by Michael Douglas in the 1987 film *Wall Street.*

Venter had his own hidden agenda. His was pushing his own candidate to head HGS: George Poste, the forward-thinking chief of R&D at the drug giant SmithKline Beecham. Venter knew Poste from their days together in Buffalo, where they both held positions at the Roswell Park Division of the State University of New York. While Venter studied his adrenaline receptor, Poste was an experimental pathologist investigating how cancer spreads.

Steinberg approached Poste about the job, but Poste wasn't about to leave his prestigious post at SmithKline less than a year after he'd secured it. So Steinberg told Venter he wanted Haseltine. Venter said he wouldn't allow it, and he reminded Steinberg that he had been promised veto power over any candidate. Venter said he didn't want the bum in there.

There was no written agreement, and Steinberg offered the job to Haseltine anyway.

But Haseltine was cagey. He was not going to jettison his Harvard parachute until HGS was on solid financial footing. He said he would resign his Harvard post if and when the company had inked a $100 million deal with a big pharmaceutical company. That would give his dream a chance of becoming real. In the meantime, Haseltine agreed to take a sabbatical to playact as CEO.

But on Steinberg's shaky $70 million promise, eight people from Venter's and Fraser's labs left their secure government jobs to join TIGR. They began collecting tissues from different parts of the body, isolating the RNA footprints of their expressed genes to build their database of human genes.

Fraser's lab group had a particular interest. At the NIH, they'd been studying an important group of body proteins called G-protein-coupled receptors that sit on the surface of cells, receiving chemical signals to give us taste, smell, and numerous other body functions. More than 60 percent of prescription drugs today are targeted to several of these receptors. (Recent examples include the receptor targeted by Claritin for treatment of allergies and the stomach receptor targeted by Zantac, Tagamet, and Pepcid for acid indigestion and ulcers.) But in 1992, only a few dozen of the G-protein-coupled receptors in the body had been identified and cloned, and Fraser's team knew there were at least ten times that many in the body. They wanted to use the EST approach to find most of the rest.

IN SEPTEMBER, THE forty-nine-year-old Haseltine went on sabbatical and moved to Washington. He and his wife, Giorgio perfume creator Gale Hayman, took an apartment at the Watergate Hotel, from which Haseltine would begin drafting his plans for HGS. Officially, Haseltine was still a scientific adviser. Harvard forbid its professors to hold any executive position in a company in which they have equity. Steinberg had given Haseltine 5 percent of HGS, which would increase to 10 percent if he become CEO.

But Haseltine already called the shots. His first step was to change their financing strategy. Steinberg had been carving up Venter's then largely theoretical genetic wares into disease categories—from heart disease to cancer—and shopping the pieces to firms interested in finding drugs for those diseases. He was on the verge of a deal with one biotech firm and was in negotiations with others. Haseltine told him to cancel the deal and all scheduled meetings or he'd quit.

When they found a gene, Haseltine argued, they wouldn't know whether it would be useful for fighting neurological disease, heart disease, cancer, or all three. It would be impossible to categorize their assets that way. What's more, he felt that the only way to get a really big chunk of money up front—the kind of capital he needed to create drugs—was to make a deal in which one big drug company gained exclusive rights to all their wares. That would also garner high royalties, which HGS would need later on to support its work.

But even before they started shopping for such a lucrative deal, they had to get their sequencing facility up and running. HGS transferred a $5 million loan it received from HealthCare Ventures to TIGR to finance the job. TIGR's sequencing facilities were operational by January, and researchers there promptly began sequencing DNA fragments from the collected tissues.

The same month, HGS opened its doors in a redbrick building in Rockville, Maryland. Haseltine had convinced his former postdoc, Craig Rosen, to head his research team, even though Rosen and his wife had just moved into a house in New Jersey, near Rosen's then employer, F. Hoffmann-La Roche. Steven Rubin, also from Roche, would help direct HGS's molecular biology group, which Haseltine staffed with a group of brilliant students from China.

These first employees got to work on the first stage of Haseltine's fifteen-year plan: to use Venter's technology to find portions of every gene in the body. HGS was already more than a licensing appendage to TIGR.

In HGS's new laboratory, biologists began collecting tissues and isolating genes to be sent over to TIGR for sequencing and analysis. Samples of human heart, liver, brain, or bone arrived at HGS after being removed from fresh corpses, tissue banks, or patients in hospital operating rooms and plunged into liquid nitrogen at a chilling minus 70 degrees. Each bit of the body was ground up and its extraneous material removed, leaving only the molecular footprints of active genes: a very fragile substance called messenger RNA (mRNA).

Like footprints in the sand, mRNA disappears quickly unless scientists chemically preserve it in another form, like a photograph. In this case, the "film" is so-called cDNA, or "copy" DNA, a stable laboratory copy of our genetic shadows. Miniature developers—bacteria—reprint each snapshot through the process of gene cloning. The bacteria accept the cDNA as their own genetic material and copy it when they reproduce themselves.

In the HGS laboratory, the bacteria multiplied in clusters or colonies: small white spots peppering a gelatinous coating on a circular plate. In each spot lay many copies of the same gene, separated now from the rest. At another station, people carefully plucked the colonies with toothpicks—a process now done by robots—and plopped them into one of ninety-six small pits or wells in a plastic plate. In these pits, the bacteria grew, reprinting each gene millions of times, with a different gene in every pit.

The idea was to isolate each gene and make enough of it to work with so that its chemical code, its string of DNA letters, could later be spelled out. That code would then be entered into the HGS gene anatomy database, a computerized catalogue of genes organized by where they are used in the body and under what circumstances.

As the plates filled with genes, technicians with gloved hands transferred some of them to seven-foot-tall green-gray freezers to create a frost-coated archive of every gene in the body.

The genes came not only from hundreds of different tissues but also from different kinds of people. Some were elderly, others children, and some not yet born. The age of the person in which a gene was found could reveal glimpses of the story of human development. Meanwhile, comparing genes harvested from healthy tissues with those from organs wracked by diseases such as cancer or Alzheimer's could yield clues to the cause of such conditions.

But Haseltine and Steinberg focused on even more directly practical benefits. In early 1993, Steinberg's investment team had begun trying to find their drug company financier, and they had begun meeting with the heads of R&D from a few big drug companies. To entice them to purchase the genetic data, they spelled out exactly what having access to that data would mean to them.

The vision they trumpeted was no less than a revolution in drug development. At pharmaceutical firms around the world, the search for a new medicine typically began with clues from academic research about what molecules played key roles in a disease. Drug company scientists often looked for big proteins stuck to the surface of cells (drug "targets") in the body that they could therapeutically tweak—block or activate—with a small chemical drug. They also had some interest in proteins circulating in the blood—such as insulin and human growth hormone—which can make good drugs in themselves.

But such disease-related molecules, the critical starting points for drug development, were then in short supply; they often emerged only after years of research into a disease. At that time, the entire pharmaceutical industry was working with fewer than a hundred drug targets.

Haseltine and Steinberg pitched a faster way to find promising protein drugs and targets. Their emerging database of human genes, with its information about where and when they were used in the body, could be intelligently combed to pinpoint quickly previously unknown proteins likely to have therapeutic value. Plugging in simple facts about a disease, such as its location in the body and the general type of molecule that might be involved, would produce a small set of promising proteins, which could then undergo laboratory testing for medically useful effects.

Most drug company executives were skeptical of, if not simply bewildered by, this newfangled approach. But SmithKline's George Poste, a visionary in the usually conservative world of big pharma, became a convert. Poste realized that HGS had a treasure trove of potential drug leads that could put SmithKline on top for years to come. He convinced his bosses to ante up $125 million for 7 percent equity in HGS and exclusive rights to all TIGR's genetic information. Though many of his colleagues thought SmithKline was crazy, Poste felt it would turn out to be the steal of the century.

HGS's investors were thrilled. The $125 million, committed in May 1993, could fund the entire $70 million commitment to TIGR, with money left over for growing the company. So far, the investors had put up only about $10 million, so they made more than ten times their money in less than a year. Steinberg was transformed into a visionary in the venture capital community.

But Venter was glum. To him, HGS and TIGR were more than an investment. They were a chance to further his grand vision. The new deal, as he saw it, deadened their impact, because all of their genetic data would be funneled to one firm rather than being widely disseminated. Other pharmaceutical companies were frozen out.

For Haseltine, though, the deal gave him the funds to build the company of his dreams. He cut his umbilical cord to Harvard and submitted his request for a leave of absence. This was a more irrevocable step in which he had to give up his grants and relinquish his chairmanship. That same month, he was anointed CEO.

Haseltine then immediately proceeded to build the first firm to use genetic information to produce new protein-based drugs. It was a risky move, and Walton, among others, warned him about it. Hundreds of biotech firms had ventured down this tortuous route of bringing drugs to market—and failed.

But Haseltine didn't really see an alternative. Even if HGS hadn't already sold all its genetic data to SmithKline, Haseltine saw the data-services business as a dead end. The customers were limited—as the information was useful to a few dozen pharmaceutical companies—and so was the pool of data. "We'd have no business in ten years," he said. From the start, Haseltine wanted to build a new pharmaceutical company, and so what if people thought he was nuts?

Before the SmithKline deal, HGS's growing cadre of researchers had been collaborating with the TIGR team in collecting tissues and isolating their active genes, but all the sequencing took place at TIGR. After the deal, HGS used its new capital to duplicate every-thing TIGR had at HGS. By November, HGS had nearly five times TIGR's sequencing capacity.

Every day, dozens of boxy sequencing machines at HGS, their screens streaked vertically with multicolored strips of red, green, yel-low, blue, spelled out the code of new genes, each color revealing one chemical letter. The machines spit out the code as fragments that were five hundred letters long. These fragments were just long enough to create a unique identifier for a gene—an EST—that could reveal key features about the gene and serve as "bait" for fishing out the complete gene. The machines produced about 750,000 genetic fragments per day, transferring millions of DNA letters into a large computer in the corner of the room.[2] This was the raw material for the gene anatomy database, to which TIGR's gene data was also added.

HGS's new competitive facilities etched a new fault line between HGS and TIGR. That schism deepened because of a fundamental conflict between the two men over the release of TIGR's data. When TIGR was established, Venter and Fraser had negotiated the right to publish their discoveries after a six-month delay, during which HGS would be able to review TIGR's data and file any patents. HGS could extend the publication delay to as long as eighteen months for the limited number of genes its researchers or SmithKline's were actively considering for possible drug development. The delay would give HGS a head start on competitors.

The contract seemed more than fair to the venture capitalists and reasonable to Venter at the time. The six-month time frame was in line with NIH guidelines for academics then. But the contract between TIGR and HGS failed to specify exactly how many gene sequences HGS could subject to the longer delay. This became a big sticking point. Fraser and Venter thought Haseltine took advantage of the vague language to hold hostage hundreds of gene sequences, including those for all the receptors that were the centerpiece of Fraser's work and the reason her lab moved to TIGR. And so little was known about most of these genes, they argued, that it would be impossible for scientists to determine any commercial use for so many of them even within the eighteen-month time frame.

The bulk of TIGR's data would be published a year or two hence, once it reached critical mass. Meanwhile, TIGR, along with HGS and SmithKline, set up a computer database in which to deposit the accumulating genetic data so that academic researchers could see it prior to publication. However, Haseltine demanded that anyone logging on to the database formally agree to give HGS first option on commercial rights to any results stemming from use of the data. Most scientists considered the terms onerous and refused to sign.

Venter's academic colleagues accused Venter and TIGR of playing a nasty game in which they hid their data for commercial gain. They felt the data were precompetitive and should be available to all for the sake of science, and medicine. Venter begged for better terms at almost every HGS board meeting. But the board, composed mostly of businessmen, was unsympathetic. After a while, the confrontations became so unpleasant that Venter sent his chief legal counsel to the meetings in his stead. TIGR researchers had to make their data available, the lawyer would say, or they'd look like idiots. But the argument fell on deaf ears no matter who made it.

Haseltine saw no sense in divulging potentially lucrative secrets until those secrets had been legally protected. He felt unfairly blamed for abiding by the rules of the agreement that underpinned the entire endeavor—and the wealth Venter himself accrued from it, having received 10 percent of HGS founding stock. When publication restrictions drew fire from colleagues, Haseltine became Venter's scapegoat. "In that debate, I got a huge black eye," the outwardly tough HGS head said years later. "Craig used to say: 'I want to give the data away free, but Bill doesn't. Bill wants to keep it private.' But he'd signed an agreement that said he couldn't do that. He made personal money; he built this institute. And then he started blaming me for the publication restrictions that he'd signed up for!"

Haseltine's legal team was filing patents on the data. By April 1, 1994, HGS had filed twenty-five patent applications on genes and clusters of partial genes, and Haseltine made it clear the filings would continue unabated. Academics felt such practices threatened their freedom to do research. Unlike Venter, however, Haseltine was unfazed by such rhetoric, which he deemed alarmist and naïve to the ways of commerce. "They didn't appreciate that this was a new way to find drugs," he recalled matter-of-factly.

But the situation was literally eating away at Venter. In 1994, Venter contracted diverticulitis, a painful inflammation of the intestine to which excessive tension is thought to contribute. By the end of the year, his condition had become so severe that it required emergency surgery to ward off an intestinal rupture.

Meanwhile, the business side of the venture was flourishing. In December, Haseltine took the company public, leading a successful road show, in which HGS officials traveled the country to pitch their stock to large investors, who bid to own the shares the company would soon sell to the public.

On Thursday morning, December 2, the company's board was slated to meet. By coincidence, HGS stock was scheduled to open on the NASDAQ stock exchange the same morning. The board convened at nine A.M., nervously anticipating the start of trading at nine-thirty.

At nine-ten, they learned that the road show had been so successful that the bankers were going to open the stock at about $14; the maximum price listed on the prospectus was $12. The board was thrilled. At nine-thirty A.M., the market opened. No shares traded. At ten-thirty A.M., no shares traded. The board was getting nervous.

An official at Smith Barney, an investment bank involved in the offering, piped in by telephone. We've got a bit of a problem, he said. We've got all buyers and no sellers. They had to raise the price.

The shares started trading shortly after ten-thirty at about $15 each. At the close of the board meeting, the share price was $21, and the following day, the stock reached its short-term high of $28. "This was an incredible party for us," Walton recalled.

Shortly after HGS went public, Venter reportedly sold his equity position in HGS, for more than $9 million in cash, part of which he would use to buy an eighty-two-foot yacht. This eliminated Venter's personal financial incentive in the success of the firm, further distancing the two organizations. One thing was certain, however. Venter and Fraser were poor academics no more.

Venter's EST technique, once scorned, was catching on. Big pharma was so upset at being denied access to EST data in the wake of the SmithKline deal that they made it known they'd welcome another source. Their cry was answered by Incyte Pharmaceuticals of Palo Alto, California, a company founded in 1991 that had been eyeing white blood cell proteins as potential therapies. Incyte jumped into the gene business by developing a TIGR/HGS–like database and broadly selling access to it. Pfizer, for example, paid Incyte $25 million for the privilege of peeking at their data.

And while academics still called Venter's work superficial, since it offered no clues to the function of genes, even they desperately sought access to the data. Academic genome researchers wanted to use Venter and Haseltine's genetic sequences to help them zero in on key disease-causing genes in the genome sequence once they unraveled it. In addition, they recognized the importance of the database by itself in speeding up research that could lead to gene-based medicine. For example, oncologists Bert Vogelstein and Kenneth Kinzler of Johns Hopkins University used the HGS-TIGR database to find three human genes associated with colon cancer. That discovery could save the lives of ten thousand people each year, Venter predicted, through earlier diagnosis of patients at risk for the disease.

By 1994, TIGR and HGS had together plucked out about thirty-five thousand human genes (in about 175,000 parts) from tissues ranging from liver to heart to brain. The data was logged into HGS's growing database, and in a partial solution to the publication dispute, Venter was finally allowed to publish it, excluding those genes HGS wanted to keep under wraps because of their potential commercial worth. The data appeared in a special supplement to *Nature* the following year.

As enthusiasm for ESTs built in the world at large, Venter's EST operation at TIGR was losing its momentum. Not only was that operation now duplicated at HGS and Incyte, but TIGR scientists also felt they were reaching a point of diminishing returns. They'd been randomly sampling active genes from about three hundred different body tissues, and, at first, every new gene snippet they sequenced was new. But after they'd sequenced most of the common genes, as they had by this time, they were much more likely to pull out a fragment they already had than they were to find something new. Sequencing the same abundant genes ten or fifteen times before finding a new one was frustrating. They felt they'd pushed this method as far as they reasonably could.

Fraser was also frustrated by the roadblocks erected by Smith-Kline's interest in her work. This was the project to use ESTs to uncover molecules called G-protein-coupled receptors that play an important role in the perception of tastes and smells, among other critical body functions. When Fraser began her project, these receptors were not hot commercial property. Receptors are stuck to the surface of cells, and while they may be targets for human-made drugs, they can't be used as drugs themselves. HGS was interested in circulating proteins that could be drugs. It wasn't set up to handle receptors.

But SmithKline Beecham was. G-protein-coupled receptors had been so historically important as drug targets—they are targets for ulcer and allergy medications, for example—that they were absolutely irresistible to SmithKline, which had first dibs on TIGR's discoveries through its agreement with HGS. Every time Fraser's group found a new G-protein-coupled receptor, SmithKline snapped it up for its drug-discovery program. All of her teams' discoveries were clearly going to be subject to the eighteen-month rule, obliterating Fraser's hopes of publishing them in a timely matter. The project hardly seemed worth pursuing.

Haseltine's scientific army soldiered on, avidly scouring tissues for genes for more than another year. By the end of 1994, Haseltine claimed that his firm had already captured a "substantial majority" of the estimated one hundred thousand or more human genes.[3]

But the team at TIGR was more than ready to latch on to a new dream. In fact, they'd already found one, in the form of a delightful and gifted, if somewhat weary, professor at Johns Hopkins. Hamilton Smith had dedicated his professional life to studying an invisible pest that was about to become Venter's next trophy and, inevitably, the source of his life's next injustice.

7

FAME IN A GERM

The same month SmithKline Beecham bought its $125 million stake in HGS, a very tall, middle-aged gentleman was slated to share the stage with a younger colleague at a genetics conference in Bilbao, Spain. Hamilton Smith, a sixty-two-year-old molecular biologist at Johns Hopkins University with a Nobel Prize to his name, spotted the younger scientist on the sidewalk outside the meeting room before the session was scheduled to begin. Smith had heard the acerbic comments that trailed J. Craig Venter—for the supposed silliness of his EST work and the evils of his attempts to patent them—but he had never actually met the man.

"Where are your horns?" Smith asked Venter with a disarming chuckle. "I heard a lot of people in academia don't think too highly of you." Venter smiled. Smith was mocking the rumors; he never put too much stock in what other people said. He would form his own opinion of Venter, just as he did everyone else.

As Venter ascended the podium inside and began to speak, Smith was impressed by Venter's energy—and by his results. Venter was clearly generating a huge amount of information from his EST sequencing work.

Venter's enthusiasm was a delight to this deliberate, excessively humble man who had grown tired of the accolades heaped upon him for his prize, which he never felt he deserved. Smith won the Nobel for his work on an obscure germ called *Haemophilus influenzae*, which, despite its name, is not related to the flu virus. It isn't a virus at all, but a bacterium, an independent single-celled creature that could infect the ear, nose, and lungs of children and sometimes cause pneumonia and meningitis.

In experimenting with this germ in 1968, shortly after Smith had arrived at Hopkins as a junior professor, one of Smith's graduate students had tried to infect the bacterial cell with a virus. But every time the student tried to do so, the virus disappeared. Smith thought the bacterium might have a secret molecular weapon that it used to destroy the virus. Soon he discovered the nature of that weapon: a tiny molecular knife that chopped up DNA, including the DNA inside the virus.

It turned out that the knife could cut DNA from any source, but it made its cuts only at certain places in the DNA molecule, places with a particular sequence of base pairs. Smith had encountered an incredibly powerful molecular tool, the first "restriction enzyme," that would enable scientists to snip out specific genes and tuck them into bacterial cells that would copy the genes many times over. Scientists used restriction enzymes to clone the first genes and to genetically engineer important protein medications such as insulin and human growth hormone. Smith's stunning insight helped launch the entire biotechnology revolution.

In 1993, fifteen years after winning the Nobel, Smith's own career appeared to be winding down, in part because he didn't have the energy to wind it up. But all that was about to change.

At the second evening of the meeting, Venter found Smith sitting alone at the hotel bar and asked him to dinner. The two men exchanged life stories and developed an instant rapport. Venter held Smith's contribution to the history of science in high esteem, and Smith was in awe of Venter's current pioneering work. Smith liked Venter's openness, too. Despite the younger man's obvious self-confidence, Smith saw nothing pretentious about Venter, a man more rebel than royalty. And Smith enjoyed Venter's penchant for speaking freely and frankly about his feelings and his life.

At the end of the evening, Venter asked Smith if he would become a member of the scientific advisory council for TIGR. He'd be among a small group of scientists who visited the institute twice a year and evaluated its projects. Smith said he'd certainly consider it, and Venter immediately followed up with a formal invitation sent to Smith's lab at Johns Hopkins.

The following September, Smith was attending his first advisory council meeting when he made the proposition that would change the course of his and Venter's lives.

"Would you be interested in sequencing a bacterial genome?" Smith ventured timidly. He was suggesting they might, instead of sequencing just random snippets of human DNA as they did with ESTs, tackle the genome of an entire organism, albeit a small one: that of his pet critter *Haemophilus*.

One of the goals of the government's Human Genome Project was to sequence the complete genomes of other simpler creatures, because doing so can help biologists figure out the functions of human genes. There are many similarities between the genetic blueprints of humans and even primitive bacteria. If researchers can determine a gene's function in a simpler creature, that function often provides clues to its role in humans. Projects to sequence the genomes of the bacterium *Escherichia coli*, baker's yeast, and a round worm, *Caenorhabditis elegans*, were under way at universities but were still far from finished.

Venter liked the idea immediately. It appealed to his sense of boldness. He gave the go-ahead and urged Smith to start preparing DNA for the project.

This wasn't a straightforward matter. The *Haemophilus* genome stretched for 1.8 million base pairs. But sequencing machines could read DNA only in short strips of a few hundred base pairs. So the DNA first had to be broken up to create a library of DNA pieces. Initially, Smith assumed he'd make two libraries. First, he'd make a library of big chunks of DNA and figure out how those chunks fit together to re-create the bacterium's genome. Then he'd break up those chunks and make a second library of small pieces that would be sequenced. In the end, the sequences of the small pieces would be reassembled to re-create the sequences of the chunks, which in turn would be assembled into the sequence of the genome. This was the same two-step strategy that government-funded researchers were using for the human genome.

With this in mind, Smith went back to his lab group at Johns Hopkins. "We have an opportunity to get the sequence of *Haemophilus*. All we have to do is construct a library and do some mapping," he said, referring to the ordering of the larger chunks of DNA. Nobody was enthusiastic. "It's too much work," they complained; "we don't have a grant for it." They were already involved in other projects. One student said simply: "What's in it for us?"

Smith trudged out of the lab, deeply annoyed at his lab group for their pessimism. He felt it was the opportunity of a lifetime. TIGR, the best sequencing laboratory in the world, was willing to sequence their organism. This would yield an incredible bounty of information that they could apply to the biological problems they'd been working on. He was determined to find a way to make the project possible, even without his own group's cooperation.

Alone in his office, he began to think. There was a potentially faster way, and one that might not require the help of his graduate students and postdocs. Perhaps he could do away with the mapping stage—the big chunks, that is—and simply shear the entire *Haemophilus* genome into the millions of small pieces in one fell swoop. After sequencing those pieces, a computer might put them together into a genome, like Humpty Dumpty reassembled after the fall.

It was an incredible wish. This method, called whole-genome shotgun sequencing, had been tried before with viruses whose genomes stretched for a mere fifty thousand base pairs. But the strategy had never been used on anything close to this length. It wasn't clear whether a computer could solve a puzzle containing so many small pieces. This was why researchers rejected this strategy for the human genome at Charles DeLisi's 1986 meeting in Santa Fe. Even scientists sequencing the *E. coli* genome, at 4.6 million base pairs, were told that shotgunning would fail because of the large number of pieces to assemble.

Sentiments had not changed since. The genetics establishment had laid down the two-step mapping strategy (involving the chunks) as the law for large projects. Smith's whole-genome shotgunning plan ran smack up against that law.

But Smith thought shotgunning was a natural approach for the TIGR team. The random spelling out of small DNA strips from a genome was very similar to the random sequencing of human gene fragments for which TIGR was famous. TIGR had even developed some software that could be used for linking together different ESTs into longer pieces of DNA representing whole genes. Smith thought the software might be adapted to link all the little pieces of *Haemophilus* DNA. But he didn't really know whether it would work. If it failed, he would look like a fool.

But if it succeeded, the project could forever change how microbial genomes were sequenced. The strategy might be used to speed up dramatically the decoding of many other disease-causing bacteria, for example, providing genetic clues scientists could use to design new and desperately needed antibiotics. It might be employed to rapidly decipher the workings of single-celled creatures that degrade toxic substances or withstand blistering heat, insights that could someday improve the efficiency of environmental cleanups or industrial processes.

Smith wrote a brief computer program that assembled short sequences to see how it might be done and to convince himself it could work. The program indicated that the vast majority of the bacterium's genome could be re-created by sequencing twenty-five thousand DNA slivers. That would amount to sequencing 12.5 million bases, or the entire genome about seven times over. TIGR had the machinery to sequence about a thousand fragments a day. If they devoted all their capacity to the project, and nothing else, they could be done in as little as a month!

The shy Smith didn't tell Venter right away. But two weeks later, Smith ran into Venter during one of his frequent visits to TIGR, and Venter asked: "How's the library coming?" Smith said that he couldn't really make the library as they had planned. "I couldn't find anyone in the lab willing to work on that," he told Venter with regret. "We don't have the time or the money." But he had another idea, and he wanted to explain it to the TIGR group.

A few weeks later, Smith stood up in front of a dozen TIGR staff members to lay out his plan. He displayed a chart depicting how the assembly might work. If they sequenced five thousand pieces, they'd be able to fit together only a relatively small percentage of the genome. If they did ten thousand, more of it would fall into place. Eventually, with twenty-five thousand pieces of sequence, the genome would assemble into a long piece of DNA that would fold back on itself to make a circle, just as the *Haemophilus* genome does.

Smith didn't have to explain that he was talking about shotgunning something almost fifty times bigger than had ever been done before using this method, or that no one had written a computer program designed to do the assembly Smith was envisioning. Smith's program did a reasonable job, but it was just a prototype. And nobody really knew whether the TIGR software Smith hoped could be used for the assembly could really handle the job.

But Smith was confident. TIGR had all the necessary tools. This was a job just waiting to be done.

There was a long, stunned silence. The stakes were huge. If they were going to do this, they'd have to do it right—and that would mean dropping just about everything else, as the job would require almost all of TIGR's machinery. Then if TIGR's only project failed, its reputation and long-term survival could be in jeopardy. The TIGR staff thought about the risk for their careers.

Venter finally broke the silence. "Let's do it," he blurted out. He loved the idea.

Once Venter chimed in, the rest of the staff quickly agreed. They would have followed Venter no matter what he'd decided. They'd learned that Venter had a superb instinct for how to proceed. However, most of the group was also genuinely thrilled by the possibility of taking this on.

In early 1994, Venter and his top advisers, including his wife, began mapping out their strategy, anticipating problems and projecting the costs. They knew they'd be testing the limits of their software, but they thought they at least now had the computer power, having recently purchased some state-of-the-art parallel computers.

At about the same time, molecular biologist Robert Fleischmann, who was leading the project, wrote a grant application to the NIH requesting a half million dollars to do the work. In the application, he roughly described how his team would assemble twenty-five thousand pieces of *Haemophilus* DNA and close the remaining gaps. He and his TIGR coauthors thought they made a strong case. If the team succeeded, making a library of big chunks of DNA might be eliminated as a step in sequencing, which would have huge implications for many sequencing projects including, perhaps, human.

Back in his lab at Hopkins, Smith immediately began pouring, shaking, and sifting vials of *Haemophilus* bacteria in an attempt to construct the necessary DNA libraries. Since nobody would help him, he made the libraries himself. It required first extracting the DNA from billions of *Haemophilus* bacteria, mechanically chopping it up by sending it through an aerosol mist, then inserting about a hundred thousand of the resulting pieces into a so-called vector that enables them to be cloned inside bacteria. For his vector, Smith used a circular piece of DNA called a plasmid that can live symbiotically inside a bacterium and that is duplicated whenever the bacterial cell divides. (Duplicating each piece of DNA was critical, because the chemical reactions that make sequencing possible won't work without numerous copies of a piece of DNA.)

On paper, the process was straightforward, but in practice it was tedious and frustrating, with all sorts of annoying snafus that slowed progress. It took Smith three months to make sure that each plasmid had exactly one piece of *Haemophilus* DNA and no more. He had to devise a kind of scale for plasmids to make sure they were the appropriate mass for a plasmid toting a single piece of DNA. In February 1994, he brought his third library—a culture of bacteria that collectively contain the various pieces of *Haemophilus* DNA—down to TIGR. They checked it, sequencing about five hundred to a thousand clones. It was 99 percent pure. In April, the TIGR team began deciphering Smith's clones in earnest.

One day later that spring, Smith realized the effort might have far more significance than anyone had first thought. He walked into Venter's office and shared with him his revelation. They were unraveling the complete genetic instructions of a living creature for the first time, he told his younger colleague. Scientists had decoded the genetic instructions of viruses, but viruses were not free living. They could not reproduce on their own.

Venter was stunned. He'd always thought of the sequencing effort as a fun technology project, something that went against the grain of how genomes were supposed to be decoded; no one else was doing a shotgun-sequencing project. But now he suddenly understood that the project was also a race to decode the first complete set of genes of a living thing. Here, there was competition.

Using the sanctioned two-step methods, Fred Blattner in Wisconsin was deciphering the twofold-larger genome of the *Escherichia coli* bacterium. After five years, Blattner's team had about a third of that genome done. The international effort to decode the huge genome of the worm *Caenorhabditis elegans* still had a ways to go. But there were rumors that other groups—such as Walter Gilbert's at Harvard and Sydney Brenner's at UC Berkeley—might be sequencing bacteria at a faster pace.

Realizing the stakes, Venter began pushing his team hard. By August, Venter's team had generated more than 80 percent of the sequences they needed. Then the "pink sheets" arrived in the mail. On them were the comments from the scientific committee that had reviewed TIGR's grant application to shotgun *Haemophilus*. The upshot: Nobody believed they could do it. Comments ranged from "overly ambitious" to "ridiculous." The project would not be funded.

❧

THE REVIEW COMMITTEE was not completely crazy to question the project. The biggest hurdle of the project still lay ahead. Could the team reassemble the twenty thousand pieces of sequence it had generated? The TIGR gang was nervous about the assembly. The first assemblies had made use of a computer program written by TIGR's young star programmer, Granger Sutton, a reserved but amiable computer scientist who had come to TIGR as a postdoctoral student in 1992.

Sutton had designed the software to link clusters of gene fragments (ESTs) into entire genes based on their common sequences. But this required tying together a hundred fragments at most. So when the researchers began using the program in the summer of 1994 to assemble the thousands of DNA slivers they had sequenced from *Haemophilus*, they ran into trouble. Not surprisingly, the program crashed as soon as the number of pieces grew to more than two hundred.

TIGR's sequencing director, Mark Adams, began hounding Sutton to write a better assembly program. At first, Sutton blew him off. He was busy with his "real" job—to improve the accessibility of the EST database—and didn't feel he had the expertise to design a good assembly program. But Adams kept after him. "Why don't you just write one?" he asked.

In September 1994, most of the TIGR staff traveled to Hilton Head Island, South Carolina, for the four-day Genome Sequencing and Analysis Conference. There, Fleischmann reported his team's progress. They had virtually finished the basic sequencing of the germ's genome and had assembled it with Sutton's original program into more than 539 pieces, which they had further reduced to 236 with additional laboratory work. The results were impressive. But in actuality, Fleischmann was not at all confident that his assembly was correct. Furthermore, everyone knew that assembling the final pieces—which contained confusing repeated DNA sequences—would be very difficult without new software.

When Fleischmann returned to Maryland, Sutton had a surprise for him. While most of his coworkers were away at the conference, Sutton had been at work in the empty TIGR offices. For four days, he'd pieced together pieces of several programs he'd already written and inserted new computer code. By the time the TIGR team came home, he was done. Sutton said: "I think I have an assembler. Where's the data?"

Sutton ran the raw pieces through his new program and voilà! The program put them together. The excitement mounted throughout the institute.[1] Save for a couple of hundred gaps, Sutton's program assembled the entire genome.

Over the fall and winter, TIGR biologists—along with two of Smith's postdoctoral fellows at Hopkins, who had belatedly decided to participate—managed to close virtually all of the remaining gaps.

One Friday in April 1995, one of TIGR's research associates handed him the data that was supposed to fill the last two gaps in the genome. Fleischmann took a hard look at the data and, to his dismay, saw that it did not appear to fill the gaps. He went home that weekend crestfallen, certain there was a mistake elsewhere in their beautifully assembled genome.

But Monday morning, Fleischmann walked into his laboratory and took another peek at the new data. Suddenly, he realized he had misunderstood what his researcher had told him on Friday. The data did match the holes it was supposed to fill. The genome had come together into a complete circle! It was finished.

In May, Venter and Smith announced their triumph in a keynote speech to an audience of thousands at a microbiology meeting held in a Washington, D.C., hotel. The TIGR group, Venter declared, had decoded the genetic instructions of the first living thing. They'd spelled out 1,830,137 base pairs by fitting together some twenty-six thousand random slivers of DNA. And they'd used a highly unorthodox strategy for DNA pieces even remotely that large. At the end of their hour-and-a-half presentation, they received a standing ovation from the audience, something virtually never seen at a scientific meeting.

In the media, the TIGR group made a splash as well. *The New York Times, Newsday, The Pittsburgh Post-Gazette,* and many other newspapers across the country carried the story. In the *Times,* Wisconsin's Blattner called it "an incredible moment in history," and Francis Collins hailed the result as "a significant milestone." In *Newsday,* a microbiologist from Vanderbilt University named Martin Blaser lauded the TIGR team's complete microbial genome as "the Holy Grail."

Biotech executives deemed the work a major breakthrough. Whole-genome shotgunning might be used to decipher the genomes of pathogens far more important than *Haemophilus,* paving the way toward new vaccines and targets for antimicrobial drugs. More directly, the work could lead to new antibiotics for treating *Haemophilus* infections in children.

The work also impressed Haseltine. While the HGS chief had not been thrilled with the *Haemophilus* project when it first started, when it was over, Haseltine demanded that Venter wait to publish his findings to give the company time to file for patents on potential medical uses derived from the bacterium's genome sequence. This much, he said, was required by TIGR's agreement with them.

Venter went ahead with publication without waiting the six months stipulated in the agreement. After all the buildup, it would be a disaster for TIGR's reputation if the paper on the *H. influenzae* genome didn't appear until the end of the year. Venter suspected Haseltine was motivated more by competitive jealousy than genuine commercial interest, and he submitted his team's manuscript to *Science*.

Haseltine insisted he was not interested in blocking publication or even delaying it. In fact, he said he was willing to relieve Venter of his six-month obligation if only the TIGR head would give him all the information HGS needed to file patents on the sequence. "What irritated me the most is that he had obviously completed the sequence, annotated it, sent the paper in for publication, but hadn't given us the annotated data," Haseltine recalled. (Venter said there was nothing in their agreement about annotated data, as that agreement was meant to apply to single genes not whole genomes.) When Venter finally gave HGS the annotated data, Haseltine said, the HGS legal team turned it around within a week, and the paper was published at the end of July.

The results were important not just from a medical perspective. They also shook up the genomics community because of the unorthodox sequencing strategy the TIGR team used. Venter felt vindicated, but the painful memories endured. "People can be so incredibly stupid," he later fumed, "and then change their minds overnight, instantly forgetting there was ever any controversy."

8

A WORM SHOWS THE WAY

Venter's most formidable competition for being king of sequencing in the early 1990s was a transatlantic duo whose specialty was worms. Robert Waterston was a slight, balding American in his late forties. He was an informal man who often dressed in sandals, a T-shirt, and torn jeans and bicycled several miles to work, even in snow. His partner was an equally down-to-earth, sandal-sporting Englishman named John Sulston. Unlike Venter, these two free spirits fit comfortably into their university communities. Waterston and Sulston were similarly impatient for progress, but both were more inclined to push for change in the academic system than to quit that system entirely.

Waterston and Sulston spent their days considering a tiny translucent worm known as *Caenorhabditis elegans*, a critter no bigger than the period at the end of this sentence. Studies of the worm had spawned key insights into the development of the brain, and its genes were thought to hold secrets to the function of muscles, digestion, reproduction, and even longevity in more complex animals. And thanks to Waterston and Sulston, investigations of *C. elegans* also brought about major progress in a more technical realm: DNA sequencing.

Waterston began his career trying to isolate the genes involved in the worm's muscle development. At Washington University School of Medicine in St. Louis in the early 1980s, while Venter studied his adrenaline receptor in Buffalo, Waterston was studying worms with genetic mutations that altered the structure of their muscles. Altogether, Waterston figured his worms had mutated versions of about fifty different muscle genes. But identifying those genes was an incredibly slow process in those days, taking a year or two per gene. It would take a lifetime to find all fifty. Like Venter, Waterston was eager to find a faster way.

In the laboratory next door, Maynard Olson was piecing together chunks of baker's yeast DNA to create a "map" of the yeast genome, some twelve million bases long. The idea was to clone large segments of the yeast genome and line them up in proper order. With these segments in hand, Olson figured that obtaining key yeast genes would be far simpler and faster.

To Waterston, this seemed like a great idea. Why not do the same for the worm? With a map of the worm's one hundred million base-pair genome, he could find his muscle genes faster and other biologists could find the worm genes they wanted much more quickly as well.

But before he could begin any such project, Waterston ran into Sulston at a meeting at Cold Spring Harbor Laboratory on Long Island. Sulston mentioned that he and his colleagues at the Medical Research Council (MRC) laboratory in Cambridge, England, were already building just the kind of map Waterston had envisioned. Figuring it wasn't worth duplicating their effort, Waterston went back to studying muscle genes.

By coincidence, Waterston was invited to the MRC on a sabbatical to continue his muscle work in 1985. There was no room for him where the main MRC *C. elegans* group was stationed, so Waterston was assigned to the small outpost where Sulston and a colleague named Alan Coulson were making their map. By then, the Englishmen had compiled a library of pieces of worm DNA and had strung about twenty-five hundred of them into seven hundred longer stretches. At this point, they were stuck, unable to make any more progress. The problem was that some crucial pieces of worm DNA were missing from their collection, like lost pieces of a jigsaw puzzle. The bacteria they'd used to make their library had refused to make pieces containing about one-fifth of the worm's DNA.

Waterston talked to his office mates about their problem over drinks at the pub and in the lab, and he became increasingly involved with the project. In the summer of 1986, Waterston briefly returned to St. Louis to take care of some laboratory business. While he was there, he stopped by Olson's laboratory and was astounded to learn that Olson's team was testing a way of cloning pieces of human DNA in yeast. Olson called his technology yeast artificial chromosomes (YACs).

Waterston reported the finding to Sulston. Both of them were excited. Their hunch: Since the yeast was closer to the worm on the evolutionary tree, it might clone pieces of worm DNA that bacteria wouldn't. Olson's invention might be the key to closing the gaps in their worm map. Sure enough, Waterston was able to clone worm DNA using YACs, and he and his English colleagues discovered that the newly cloned pieces bridged many of the gaps.

As Sulston and Waterston began spreading word of their success with the worm map, colleagues who had at first scorned their efforts as a waste of time began to come around. In early 1988, the National Academy of Sciences report on the genome project indirectly endorsed their effort. It noted that understanding the genomes of simpler organisms was critical for human genetics. Since genes have remained surprisingly similar throughout evolution, the report's authors reasoned, knowing the role of a gene in one species would shed light on its role in another.

The committee did not foresee, however, the impact of Sulston's and Waterston's efforts on the Human Genome Project. But then, few people appreciated the ingenuity or verve of Sulston and Waterston.

By winter 1989, the St. Louis and Cambridge teams had put the finishing touches on their map, but they had no plans to decipher the worm's genome letter by letter. They figured somebody else would perform this tedious technical task, which wasn't their area of expertise.

But when James Watson and his advisers penned a list of the organisms whose genomes would be sequenced as part of the Human Genome Project, the worm was conspicuously missing. The list included only *Escherichia coli*, baker's yeast, the fruit fly, and the mouse. When Sulston and Waterston heard this, they felt they had to do something. They decided to volunteer to spell out the *C. elegans* genome.

They took advantage of a meeting of worm researchers at Cold Spring Harbor Laboratory, where Watson was director, to get Watson's attention and make their pitch. With $100 million, they'd have the sequence done by the year 2000, Sulston promised Watson, whom the duo had cornered in his office one evening during the conference. That was $1 a base—a bargain in a day when costs ran as high as $10 per base. Neither Sulston nor Waterston were sequencers, but both had heard about the fancy automated machines that Craig Venter was testing at the time, and they figured they could make the machines work, too. Watson was intrigued, but he wasn't about to commit that kind of money then and there.

Nevertheless, the transatlantic duo had made an impression. Waterston won a $2.25 million NIH grant for a three-year starter effort, to begin in 1990. Sulston and Coulson received the same amount, from the NIH and England's Medical Research Council. Together, the researchers aimed to sequence increasing amounts of the genome each year, reaching an unprecedented one million bases per year in the third year. It was an extremely bold plan, and Sulston's and Waterston's colleagues doubted they could do it. It seemed even more unlikely that they could ever decode all one hundred million bases.

Critics also pooh-poohed sequencing the worm as mindless, big science, not real biology. "You've got a good career," one of Waterston's friends warned him one day. "Why are you throwing it all away?" When Sulston was asked why he bothered, he joked that he had "a weakness for grandiose meaningless projects."[1]

If their goals broke precedent, their methods would have to as well. Sequencing had been a task done by small numbers of graduate students, but that wouldn't do for such a big project. Sulston and Waterston set out to create a sequencing factory—a smart, modular assembly line for deciphering worm DNA.

The two men went on a major recruiting spree, hiring people to help lead the sequencing effort and to design and run the computer systems that would organize and interpret the flow of DNA data from their sequencing machines. They looked for people who, as Sulston put it, "were skilled, cheerful and would get on with it," rather than those who were necessarily trained laboratory technicians. They wanted employees who didn't mind doing repetitive tasks, as there were plenty of those kinds of jobs on a sequencing assembly line. Sulston ended up with sixteen people, and Waterston had six on his initial team.

Each team bought two of the new automated sequencers. Combined, their sequencing power equaled that of Venter's NIH laboratory. By 1992, their project was moving along at a clip. They were on track to meet their goal of one million bases easily.

To decode the entire genome, however, Sulston and Waterston needed a massive infusion of funds. But when they felt out their usual benefactors, the NIH and Britain's MRC, they got a lukewarm response. Worried, they started looking for other patrons.

Waterston soon secured an invitation to join a new start-up company in Seattle, funded by an American venture capitalist named Frederick Bourke. Both worm specialists flew to Seattle to meet with Bourke. Though the details were vague, Bourke seemed serious about building a sequencing factory that would decipher large parts of the worm and human genomes, patent important disease genes, and spin those patents into profits.

Talks broke down over how the firm was going to make money without keeping the human genome secret. But Bourke's overture had the desired effect. It got James Watson's attention. Watson certainly didn't want to lose two of his best researchers to the private sector. He contacted his counterparts at the MRC, who in turn approached the head of the giant British medical charity, the Wellcome Trust.

The Wellcome was flush with cash after the recent sale of some of its stock, so it was a perfect time for it to find a new outlet for its wealth. It agreed to build Sulston a large new genome campus, called the Sanger Centre, at which he could sequence the worm and eventually the human genome as well. With British funding secure, the NIH finally found the money Waterston needed to complete his portion of the worm genome.

9

THE INSIDER

Bringing Sulston and Waterston back into the fold was James Watson's last major act as genome project chief.

NIH director Bernadine Healy had been dissatisfied with Watson for some time. She strongly disagreed with Watson on the issue of patenting Venter's gene fragments and was furious at him for ridiculing Venter, one of the NIH's rising stars. Healy also felt that Watson's part-time appointment at the NIH was inadequate for him to do his job well.

After the negotiations with Sulston and Waterston fell apart, Bourke complained vigorously to Healy that Watson, supposedly a public servant, was instead working against private enterprise by trying to squash his new company before it could get off the ground. Healy began an inquiry and became convinced that Watson had outside interests that conflicted with being a director of a major NIH program.[1]

Watson denied any conflicts of interest, including financial ones. But his long-running feud with Healy caused him to resign as Human Genome Project director in April 1992. He returned to Cold Spring Harbor Laboratories full-time. So as Sulston and Waterston expanded their worm project, Healy began casting around for a new director of the Human Genome Project. Her search committee kept bringing up one name: Francis Collins.

Collins's standing in the biomedical community had continued to rise since the cystic fibrosis triumph. In 1990, his group cloned the gene for another genetic disorder, neurofibromatosis, in which tumors grow along nerves and other tissues, often leading to bone and skin abnormalities or learning disabilities. By 1992, Collins's team was closing in on the Huntington's disease gene, whose general location had been mapped nearly nine years before. Once populated with just a few people, Collins's laboratory now swirled with thirty students and researchers.

Healy needed somebody who would be respected by NIH scientists. Collins fit that bill. He was popular among his NIH-funded colleagues in the genome project because he was friendly, unpretentious, and unlikely to challenge the status quo. Collins also represented the pinnacle of success of small-scale biology.

Healy also wanted a strong advocate for the project, and Collins promised to be that, too. He had stage presence, a gift for explaining science in simple terms, and a doctor's appreciation for the genome project's health and medical applications. Collins was also sensitive to the project's ethical and social dimensions, another important concern of Healy's. An open and devout Christian, the forty-two-year-old scientist had a deep sense of morality and was intensely interested in ethical issues. He'd publicly spoken about his religious beliefs and how they bolstered his commitment to biomedical research.

There was only one problem with Collins. He didn't want the job. When Healy called him to offer it in spring 1992, Collins was stunned and flattered, but he turned her down immediately. The idea of being a federal administrator did not appeal to him. And why should it? He was extremely happy at the University of Michigan. He was admired, well funded, and surrounded by colleagues he liked and respected. The NIH, by contrast, offered a low salary and lots of government regulations.

His rejection made Healy want Collins even more. She made the job as attractive as she could, promising Collins a large laboratory space in a new building—space that had already been promised to another institute years before. She told Collins he'd be leading the new genome research program on the NIH campus she planned to start. She also pointed out, again and again, that this job meant an unprecedented opportunity to advance the field of molecular genetics and to shape national science policy.

Healy eventually convinced Collins to visit the NIH in summer of 1992 to talk to her, Watson, and the search committee. Although Collins recognized the historic opportunity, he could not see himself doing it. He turned her down again.

Healy was persistent. She met with Collins's closest professional colleagues, including geneticist Mary-Claire King, and asked them to pressure him. She kept calling Collins, reaching him wherever she could, home or lab. "Francis was driven by his dreams," Healy recalled. "When that became apparent to me, I knew that sooner or later I'd be able to recruit him."

While Healy courted Collins, Collins was beginning a courtship closer to home. His long marriage to his high school sweetheart had recently ended, putting him back on the dating scene for the first time since his teens. For years, he'd worked in the clinic with a young, blond genetic counselor named Diane Baker. He was attracted to her and wanted to ask her out. But he was terrified. He got Baker's attention at a lab party, where Baker noticed him eyeing her. Baker asked Mitchell Drumm, Collins's graduate student and also a friend of hers, whether her instincts were correct.

Unbeknownst to Baker, Collins had also asked Drumm whether Baker liked him. Drumm encouraged Collins, and Collins soon mustered up the courage to call Baker at home. But when he got her voice mail, he panicked and hung up without leaving a message. Two days later, Baker invited him to a basketball game. It was the start of a six-year courtship that would lead to marriage.

Amid all this, Collins was having second thoughts about saying no to Healy. Are you nuts? he asked himself. This is probably the most exciting opportunity in all of science, and you turned it down because you are comfortable doing what you're doing?

In late fall, Collins visited the NIH to serve on a grant review committee, and Healy coaxed him into her office one more time. She had already been working on Collins for more than sixth months, but she had one last line to deliver. "I would hate to think that someday you and I are going to be in the same nursing home," she said, "and I'll be coming down the hallway with my walker and you'll be coming the other way with your walker. As we pass, you'll turn to me and say, 'Dammit, Bernie, I should have taken that job.'" It was a fanciful thought, but it tapped into Collins's biggest fear—that he would later regret his decision.

A week later, he took the job.

Collins left Michigan in April 1993, with a literal swan song he had composed himself. He stood in front of his last genetics class, whipped out a guitar, and began to sing a satirical rendition of Frank Sinatra's "My Way." With a lab coat slung over his lanky frame, inched even higher by the heels of his cowboy boots, Collins wailed his way through the last stanza.[2]

> *I'm just a man, what can I do?*
> *Open your books, read chapter two!*
> *And if it seems a bit routine, don't talk to me, go see the Dean.*
> *Just start today, love DNA*
> *And do it myyyy waaaaaaay!*

The class erupted in cheers and laughter. But the song marked a bittersweet transition for Collins. He was looking forward to his new adventure, but he was also sad about leaving his productive Michigan years behind. He'd be commuting between Ann Arbor and Bethesda to keep his nascent romance alive.

Collins had been a motorcycle fan for most of his life. He had begun sticking decals on one of his helmets to signify every gene his team had tracked down. Just before he left for Maryland, the helmet was due for a new sticker. The Huntington's Disease Collaborative Research Group, of which Collins's team was a part, found the Huntington's disease gene. The news made national headlines. One reporter called it "the longest and most frustrating search in the annals of molecular biology."[3]

❦

As soon as Collins arrived at the NIH in 1993, he faced a pressing problem. Genome project scientists had become bogged down by the first step of the two-step strategy they planned to use for sequencing human DNA. Researchers were struggling to make a good library of big chunks of human DNA, the ones that they later intended to shatter into smaller pieces that they could sequence.[4]

The two main technologies available for building this library of chunks both had major drawbacks. Researchers at the Department of Energy had been using something called cosmids to create a library of overlapping chunks from human chromosomes 16 and 19. They hoped to string together the chunks to re-create each chromosome.

But cosmids, which are pieces of bacterial DNA that include copying instructions for a bacterium, only enabled bacteria to copy chunks of DNA about forty thousand base pairs long. Using chunks of that size, one would have to overlap and order probably one hundred thousand or more of them to cover the entire human genome, a very time-consuming and expensive proposition. Cosmids also refused to copy certain chunks of human DNA and, even worse, they often scrambled the DNA, creating strips of DNA that did not actually exist in nature.

The other possible technology was Maynard Olson's yeast artificial chromosomes. YACs enabled yeast cells to carry chunks of human DNA from several hundred thousand to a million bases long. Using YACs, less than ten thousand pieces (instead of a hundred thousand with cosmids) would cover the human genome.[5] Soon after Collins assumed the helm, a team of French researchers led by a burly geneticist named Daniel Cohen was about to publish a letter in *Nature* describing a complete genome map consisting of big chunks of DNA that had been copied in YACs and loosely ordered across the genome. The buildup to this map had been intense. U.S. biologists had been very worried that a map created in France rather than the United States could end up leading to the complete human genome sequence.

But by the time the French map came out, researchers had discovered that YACs also did not faithfully copy the pieces of DNA inserted in them. They were even worse than cosmids. Yeast cells would habitually rearrange and truncate the human DNA inside them. Even more troublesome was that they would often join together two or more unrelated pieces of DNA during the cloning process.[6] Cohen's map, it turned out, was full of such chimeras, which made assembling the genome almost impossible.

Without any good human DNA libraries in hand, or even any technology to create them, Collins backed another kind of human genome map. The idea for this map also came from Olson. Since the technology did not yet exist to decipher the human genome in all its detail, Olson instead proposed creating a map consisting of small DNA signposts scattered roughly every one hundred thousand to two hundred thousand DNA bases across the human genome. If the human genome stretched from St. Louis to Los Angeles, these unique signposts—short stretches of DNA that Olson dubbed "sequence tagged sites" or STSs—would be about two football fields apart, with uncharted territory in between.

When researchers finally made a good collection of human DNA chunks, Olson argued, this map of signposts would accelerate the task of putting the chunks in order. Researchers could locate the STS signposts on each chunk and align those chunks according to the map of signposts. The signposts would also enable researchers to compare all the disparate existing maps of the genome and check their accuracy.

In a 1990 *Science* article, Olson demonstrated how his idea might be implemented. He started with a small unordered collection of DNA chunks near the cystic fibrosis gene on human chromosome 7 along with a set of STS signposts from the same area whose order and precise location were also unknown. By using a bootstrap procedure in which he determined which signposts were on which chunks, he was able to determine the order of both the chunks and the signposts.[7]

It was a sexy idea that caught on quickly. By the early 1990s, a number of U.S. teams were racing to produce signpost maps of various human chromosomes. But what most excited Collins was an even bigger effort under way at the newly minted genome center at MIT's

Whitehead Institute. There, a team of scientists led by the ambitious and politically savvy Eric Lander was building an STS signpost map of the entire human genome. It was the first attempt to take Olson's idea to its ultimate conclusion. Using innovative biochemistry and new methods for automating it, Lander's team aimed to plot some fifteen thousand of these points onto the human genome.

When the map was finally published in *Science*, the media hailed it as a giant step toward deciphering the human genetic blueprint. "When history looks back on this particular era, it will remember this day," Collins told *The Wall Street Journal*.[8] But in reality, the Human Genome Project did not have the technology even to begin serious human sequencing.[9]

Amid the hype over the STS signposts, few researchers even noticed a promising new technology under development at the California Institute of Technology. Caltech molecular biologist Melvin Simon and microbiologist Hiroaki Shizuya had deleted the DNA-scrambling machinery in *E. coli* bacteria so that the bacteria would make a single faithful copy of any piece of DNA they carried. Their manipulations enabled the bacteria to tote a piece of human DNA three to five times larger than a cosmid (though still smaller than the highest-capacity YAC). Simon and Shizuya called their newly engineered molecules bacterial artificial chromosomes (BACs).

Neither researcher did much to promote the new technique, however. Shizuya, a perfectionist, spent years honing the method before publishing the first paper on it in September 1992. Even then, the work was overshadowed by YACs, whose proponents were loath to abandon them even after they proved troublesome.

But at Lawrence Livermore, a six-foot-two-inch Dutchman named Pieter de Jong had been following the progress of the BAC technique for years through presentations Shizuya made at scientific meetings. De Jong had helped construct cosmid collections for the DOE, but he had become frustrated by his progress. His instincts told him that BACs might solve the problems scientists were having making gene libraries.

Without telling his boss, he sent two students out to Shizuya's lab to learn the new technique. BACs, he thought, were what everyone had been looking for.

De Jong parlayed Shizuya's BAC recipes into his own trademark technology, which he called a PAC. After moving from Livermore to Roswell Park Cancer Institute in Buffalo in 1993, de Jong concentrated on duplicating and distributing his PAC libraries as rapidly as possible.

PACs became extremely popular. De Jong shipped his PAC libraries to most of the NIH-funded centers and to the Sanger Centre in England. Sanger Centre biologists used PACs and BACs to link together pieces of DNA covering human chromosome 22. But as late as 1996, researchers were using a smattering of technologies, including PACs, BACs, and cosmids, and they still hadn't settled on a single library of DNA pieces that they would use as fodder for sequencing the human genome.

Collins wasn't an expert in such esoteric technology, and it wasn't his nature to force a decision on a community not ready to make one on its own. Lacking a consensus on which technology to use, Collins and his advisory team postponed a decision indefinitely.

The lack of planning would later come back to haunt them.

10

OF WORMS AND MEN

By 1994, Sulston and Waterston were becoming increasingly confident that the human genome was within reach. Though their work on the worm was far from finished, it was going well. That year, they published 2.2 million bases of worm sequence in *Nature*, much more data than they expected to have by this date. They'd developed an industrial approach that was already churning out DNA sequence at an unprecedented rate. It far exceeded the output of other labs. They thought it could be expanded to meet the needs of sequencing the human genome, about thirty times larger than the worm.

When Waterston visted the one-year-old Sanger Centre that summer, he and Sulston had a long talk. Sulston realized he had a tremendous opportunity. He'd been given a big building, and he and Waterston were the world's leading sequencers. What were they going to do? Were they going to do the human genome? For the first time, the two worm researchers seriously broached the topic of spelling out the entire human genetic code.

In this, oddly, they were alone.

Collins still felt that it was too soon to make a serious assault on the human genome sequence. The technological advances he hoped would make sequencing less costly and more efficient had yet to arrive. No one had a library of ordered human DNA pieces that could serve as a starting point for sequencing, the type of map Waterston and Sulston had created for the worm. After the failure of YACs, there was no proven technology ideal for creating a library of human DNA pieces. As a result, nearly a half decade after the dawn of the project, less than 1 percent of the human genetic code had been deciphered.[1]

On the long flight home to St. Louis, Waterston did some arithmetic on his watch, which had a built-in calculator. The results confirmed his view that it was more doable than Human Genome Project leaders thought. Over phone and e-mail, he and Sulston sketched a plan. A revolution in technology was not necessary. Waterston's team had doubled its output every year for the past four years, and Waterston expected such improvements to continue.

The sketchiest part was the DNA library. Waterston thought they could start small, constructing ordered chunks of DNA for one chromosome at a time as sequencing progressed. He recognized that there was still no ideal technology for cloning the human DNA, but he felt that cosmids would work for now and that various new technologies on the horizon were sure to be ready soon. Unlike his colleagues, Waterston had faith that technology would improve quickly.

If the project could get started in earnest, Sulston said several months later in a speech, "why fiddle around?" The *C. elegans* sequence Sulston and Waterston had so far generated had already transformed genetics research. The sooner humanity had decoded the entire text of the human genome, they reasoned, the sooner researchers could use it to come up with breakthrough drugs and diagnostic tests. It would also be cost effective in the long run. Hunting down the gene for cystic fibrosis cost an estimated $50 million. If sequencing the human genome could eliminate the need for just a dozen or so more costly gene hunts, it would more than pay for itself. The more time passed, the more money would be wasted, Sulston and Waterston felt.

It was time to alert the leadership.

In November, the duo flew to the NIH to present their plan to Collins in his office. They suggested that their two labs, which ran their machines around the clock factory-style, along with perhaps one other team, could produce a draft of the text of the book of life in five years. The catch: This first draft would include small gaps, to be closed later. With this plan, the genetic data would get out to the scientific community that much faster, they argued.

Collins seemed enthusiastic. It was the first time he'd heard anyone seriously propose a way to get the human sequencing done quickly. And Waterston and Sulston did have an impressive track record, having done more sequencing than anyone else.

But Collins was not ready to sign on. Always the conciliator, he wanted to first check with other genome colleagues to see what they thought. Waterston was scheduled to present the plan to the twenty leaders of the top U.S. genome labs at a strategy session in Reston, Virginia, the following month. They would discuss it further then.

In Reston, Waterston brashly proposed that he and Sulston could lead a full-scale attempt to spell out the human genetic blueprint, tackling at least two-thirds of the job. With a third lab contributing at the same rate, he predicted the Human Genome Project could be 99 percent done by 2001, four years ahead of schedule. By slightly relaxing the standards for accuracy—from 99.99 to 99.9 percent—the job could be done for just $300 to $400 million, an affordable sum for a project with a $3 billion budget.[2]

The plan went over like a lead balloon. It would entail sucking funds from scores of loosely related projects under the genome project umbrella—from hunting down disease genes to mapping STS signposts—to do what the genome project was originally created to do. It would be a huge shift of focus at a time when the government was doling out close to $40 million a year for genetic mapping projects, an amount that vastly exceeded its funding for sequencing. Sulston, Waterston, and one other lab leader would be big winners. Most everyone else in the room would presumably be losers.

Waterston's colleagues were irritated that he was trying to upstage them by suggesting that he, Sulston, and just one other lab could handle this next and most prestigious stage of the project. Those who had dedicated their work to making STS signpost maps of the human genome had expected eventually to turn their maps into sequence. What right did a couple of worm researchers have to take over the job? Other sequencers weren't too pleased either at the presumption that Waterston and Sulston's system was the one to use. "Those guys were just hiring more people and doing more stuff and saying, 'We've found the model; this is how you do it,' " recalled Richard Gibbs at Baylor College of Medicine in Houston. "It was a little self-aggrandizing."

Sulston was also scheduled to make his pitch to his benefactors in England. Waterston planned to come along for moral support. But the Saturday before the meeting, a car plowed into Sulston as he was riding his motorcycle to work. The blow knocked him unconscious and smashed his pelvis into several pieces. When he woke up that afternoon, Sulston's first words to his doctors were: "I have to call Bob," referring to Waterston. He said he had to attend an important international meeting in London in a few days. The doctors shook their heads. There was no way. By Monday, Sulston had made such a complete nuisance of himself that the doctors relented and said he could go if he passed a mobility test. Sulston called Waterston from his hospital room and told him to hop on a plane.

Waterston told Sulston he was certifiable. But Sulston insisted. So Waterston met his friend in Addenbrooke's Hospital in Cambridge. By then, Sulston had passed his mobility test, so the hospital staff arranged for a car to take him and Waterston to London. Sitting in his wheelchair, Sulston gave a masterful presentation to the Wellcome Trust, his persuasive powers undoubtedly magnified by his current plight.

Unlike the NIH, the Wellcome Trust didn't have any other genetics laboratories with which to curry favor. For the Wellcome officials, it came down to money. They had to decide whether they should promise Sulston all of the additional £147 million ($221 million)—to be given out over seven years—that he was asking for.[3]

The decision faced by Collins was more difficult. On the one hand, he could hardly ignore Waterston and Sulston's proposal given the goals of the genome project; he didn't want to be seen as fiddling around. On the other hand, it wasn't fair, or politically viable, to hand over the genome project to just a few laboratories, leaving the rest behind. Nor did it seem right to rush into a single sequencing strategy when there wasn't a gold standard yet.

By early 1995, both British and American leaders were angling for a compromise. That November, Britain's Wellcome Trust committed half the amount of money Sulston wanted—£70 million ($105 million) over seven years—for Sulston's group to sequence one-sixth of the genome, instead of the one-third Sulston had hoped to complete.

Several months later, the NIH followed suit. It spread out funds totaling $20 million to a half dozen laboratories for the first year of a three-year project to develop the technologies and strategies needed to begin rapid sequencing in 1999. The NIH also funded four other laboratories for human sequencing around that time. The DOE kicked in with a small sequencing effort on its own based at its Livermore, Los Alamos, and Berkeley genome centers.

The U.S. goal was to spell out a mere 3 percent of the human genome by 1999, making the sequence as accurate as possible. Its goals for sequencing were so modest in part because no one other than Waterston's team was ready to sequence on an industrial scale. Instead of moving ahead rapidly, Collins's idea was apparently to give other researchers time to catch up with Waterston and Sulston.

Waterston's lab, which received by far the largest NIH grant ($6.7 million), was the only NIH-funded group doing sequencing of any magnitude. Besides Waterston's team, only two of the funded groups—Mark Adams's at The Institute for Genomic Research and Gibbs's at Baylor—had sequencing experience. Half the grantees had been working on other kinds of projects, such as developing STS signpost maps, and had to start sequencing from scratch.[4]

Richard Wilson, Waterston's right-hand man, was astounded by the choices. He figured the development of glamorous techniques rather than smart applications of existing technology must have swayed the decision makers.

In the end, Collins and his team were swayed by a scientific community that not only was pushing them to spread out their funds, but also was not in any hurry. Few biologists then saw an urgent need to obtain the human genome sequence, as few foresaw the impact it would have on their work. Some still considered deciphering whole genomes a waste of time.

Waterston argued in *Science* magazine that the focus on accuracy over speed "is going to raise costs and slow work down a little."[5] Privately, he felt that all the foot dragging, however democratic and well meaning, was a big mistake, a fateful slowing of things. Sulston, too, was sorry his lab couldn't "take off then and there." He felt in his bones that being too conservative was dangerous. He had always thought that if you could do something, you should go for it.

Collins's leadership style struck some as akin to that of Britain's General Bernard Montgomery, who prepared carefully for every attack during World War II to maximize the chances of success. No matter if that strategy gave the enemy more time to prepare its defenses. Craig Venter, by contrast, was more like the feisty American general George Patton, who raced ahead with his troops to wage risky battles even after being told to hold back. One of Patton's dictums: "A good solution applied with vigor now is better than a perfect solution applied ten minutes later."

Ultimately, there would be costs associated with being conservative. It wasn't just the lost scientific opportunity. It was the door left open to the private sector, which was hot on just about everything related to human genes. Scores of genome scientists were cutting deals with venture capitalists to form new companies, and genome-related companies were sprouting up practically every month.[6] Genome project biologists recognized the role of the private sector in creating new drugs or developing new devices to speed research. It was when companies made fundamental genetic discoveries that academic scientists felt threatened. These discoveries not only encroached on their territory, but also could tie up the most basic genome data in intellectual property rights, as companies applied for patents on that data.

Very shortly before Waterston and Sulston came out with their plan, Collins had had to curtail the damage wrought by the defection of Venter from the NIH. The huge amount of valuable EST data he and Haseltine had created was now out of the scientific community's reach. Human Genome Sciences attached conditions to sharing that data that many academics found unacceptable, forcing them to give HGS the first option on commercial rights to any genes found as a result, for example. The problem was solved only after Merck volunteered to give Washington University money to build a similar database. Unlike the HGS database, the Merck Gene Index was available without strings to anyone who wished to access it.

But in the glow of the Merck triumph, Collins and his crew seemed to forget its main lesson: The Venter debacle would never have come to pass had NIH officials recognized the importance of Venter's ideas and funded them. With the booming biotech interest in genomics, the Human Genome Project had to move fast to discourage industry from stepping on the public project's turf—and attempting to lock up the basic genetic data critical to academic research.

The pinnacle of basic genetic data was, of course, the complete sequence of the human genome.

11

AFFLICTION IN ICELAND

As academic researchers put into action their plans to spell out the human genetic blueprint, a wild fantasy coalesced in the mind of a forty-six-year-old Icelander living in the United States. Spurred by the same advancements in gene-sleuthing technology that had ripened the government's plan, the towering, muscular Harvard neurologist began, in late 1995, to dream up his own brand of genome project.

In the dream, Kári Stefánsson would journey back to his native Iceland. There he would build a fortress where top scientists would sift through the blood of Icelanders, all 275,000 of them, to track down the genetic roots of humankind's most common afflictions. His team would pinpoint the tiny strips of genetic code that lent a hand in the development of diabetes, heart disease, manic depression, and dozens of other conditions. He would bring new prescriptions for health and cures for ailments for which modern medicine had little to offer. Stefánsson would become a hero on the tiny island of Iceland and throughout the world.

An incurable romantic, Stefánsson held fast to his fantasy. The intense, charismatic man with short tousled hair and a salt-and-pepper beard was especially taken with the idea that he might really make a difference in humankind's struggle against disease. Little did he know that he would be struggling against much more than that.

When he first moved to the United States some twenty years before to become a medical resident at the University of Chicago, Stefánsson encountered patients dying from the devastating paralysis brought on by Lou Gehrig's disease. He watched their muscles wither, their legs hang limply in wheelchairs, their heads sag, and their breath become increasingly labored before being sucked away completely. The patients' minds and senses stayed relatively intact while their bodies fell to pieces.

Stefánsson was charged with examining the bodies of these men and women after they died, looking for clues to the cause of their affliction. He came up largely empty. He could never really tell their loved ones why.

Back in Iceland, Stefánsson's older brother suffered from schizophrenia, a disorder of insane imaginings that often brings about nonsensical and sometimes destructive actions. Other scourges had taken the lives of his mother at age sixty-two and his father, Stefán, at sixty-seven.

To avoid a similar fate, Stefánsson heaved barbells in the gym, thickening his biceps, and raced around a basketball court on a regular basis. He avoided tobacco, opting for Diet Coke and chewing gum instead. He kept his lovely, blond Icelandic wife, Vala, informed about the latest medical lore so she could provide the best care for their three children whenever they were sick.

But professionally, fighting disease was more complicated. Later in his tenure at the University of Chicago and then at Harvard, Stefánsson focused his research on multiple sclerosis (MS), having switched from Lou Gehrig's disease. He and his then graduate student, Jeffrey Gulcher, spent years plucking out proteins in MS-afflicted brain tissue. Anything that appeared in the diseased tissue but not in healthy tissue was a possible contributor to the disease. But there were hundreds of possibilities, and many of the aberrant molecules Stefánsson and Gulcher found turned out not to cause the disease, but were part of the body's reaction to it—just like the body's inflammatory reaction to a bump or bruise. Determining which proteins actually triggered the disease became very frustrating.

When Stefánsson moved to Harvard in 1993, taking Gulcher with him, he decided to switch tacks. Instead of combing diseased cells for aberrant proteins, he'd look for changes in the genetic code itself that might augur disease. He'd study related people to find the specific DNA quirks that had been inherited whenever the disease appeared. Those would point to the true cause of the disease.

But Stefánsson and Gulcher were nervous. Their approach was similar to the triumphant search for the cystic fibrosis gene, which was completed six years before. However, their task was far more difficult, because MS was much more complicated than cystic fibrosis. A defect in a single gene always caused cystic fibrosis. By contrast, a complex mixture of genetic and environmental factors undoubtedly caused MS. The techniques that had worked so spectacularly well for Collins had generally failed in other complex diseases so far.

Still, Stefánsson forged ahead. Though he knew MS patients in Boston, he wanted to study bigger families than he thought he could find anywhere near his Harvard lab to give himself the best shot at homing in on the MS gene. The only place he knew that had the huge families he needed was his home country, Iceland, which had careful genealogical records that spanned centuries. He figured that such records could be used to build large family trees with people afflicted with multiple sclerosis.

Stefánsson arranged to collaborate with an Icelandic neurologist named John Benedickz who treated more than 90 percent of the MS patients in the country and who had access to small genealogical databases that might be used to start assembling family records. In the summer of 1994, he and Gulcher flew to Iceland, drew blood from Benedickz's patients, and returned with the vials to Boston.

They analyzed the blood, and Benedickz and his Icelandic colleagues laboriously assembled the patient pedigrees from lists of their ancestors. It soon became clear that Stefánsson's hunch had been right. They could use genealogy to find familial ties between patients who did not know they were related. They felt that these extended families were going to be a huge help in finding the gene.

Unfortunately, others weren't convinced. When they applied to the NIH for funding, they were turned down. Their approach wouldn't work, the critics said, because big extended families couldn't lead them to genes for complex diseases. The prevailing orthodoxy favored lots of little families, pairs of siblings who shared the disorder. Iceland was too small to have enough of those.

They twice reapplied for funding—and got the same response. Stefánsson hated to lose. So by late 1995, he was ready to give up on the NIH and move his work to Iceland. There, he'd expand his efforts beyond MS to unraveling the genetic underpinnings of many common disorders.

The idea remained trapped inside Stefánsson's head until one fall day in 1995, when he confided in Gulcher over cappuccino at a Starbucks on the first floor of Harvard's Beth Israel Hospital. Stefánsson told his friend and former graduate student that Iceland could be incredibly fertile ground for finding disease genes on a large scale.

He argued that Icelanders would be a bountiful resource not only because of their genealogical records, but also because of their genetic similarities. Icelanders all descended from a small number of settlers eleven hundred years ago, and they have remained largely isolated. As a result, common diseases in Iceland are likely to have a common genetic cause—genetic quirks shared by most of today's afflicted Icelanders because they were inherited from a single ancestor. By contrast, in a diverse population like that of the United States, the same disease is caused by different factors in different people, making any one of those factors very difficult to link to the disease.

Despite Iceland's relative isolation, Stefánsson believed that the genes that underlay diseases in Icelanders were among those that caused disease elsewhere. For one thing, Iceland's original settlers were Norwegians and Britons, who contributed a mixture of northern European genes. For another, the incidence of major diseases such as heart disease and cancer was about the same in Iceland as it was in the rest of the industrialized world. Thus, he figured Icelanders probably represented a central ancestral branch of the human evolutionary tree rather than a peripheral twig.[1] Stefánsson wanted to set up an independent company or nonprofit institute in Iceland.

Gulcher, an American from Ohio, was horrified. He didn't want to move to a place he viewed as a rock in the middle of the Atlantic. Couldn't they form the company in Boston? But Stefánsson was adamant. The enterprise had to be built in the community that supplied it with blood. It was not only more convenient; it was the morally correct way to go.

In the past, similar conversations about pathbreaking biotech ventures had undoubtedly taken place very near where Stefánsson sipped his cappuccino. This was the community that had bred William Haseltine's interest in biotech, and that of a line of Harvard biologists before him.

What Stefánsson had in mind was different from what Haseltine hawked at HGS, of course. HGS specialized in genomics, the study of huge catalogues of genes only loosely tied to disease. By contrast, Stefánsson was interested in using genetics, the study of how genes and traits are passed from one generation to the next, to link segments of genetic material to disease.

Friends thought Stefánsson was crazy even to think about leaving the safe, prestigious confines of Harvard. Few, if any, Icelanders had ever become full professors at Harvard Medical School. Why would he give that up?

The answer was as simple, and complex, as Stefánsson himself. On a simple level, if Stefánsson left Harvard, he'd have the opportunity to do better science. But taking this risk also just felt right to Stefánsson. He believed that risks were unavoidable in life—research funding in academia is uncertain, for instance, as was the birth of exciting ideas—so he preferred to choose his own risks. "The idea that when you get up in the morning you are not taking a risk is totally incorrect," he reflected later. "There is no stability in life; why don't we enjoy the ride?"

Stefánsson believed life was fundamentally shaped by big upheavals rather than by gentle changes. He thought that chaos provided the most important energy for creative work, including science. It was time for another upheaval.

In spring 1996, just as the Human Genome Project's initial sequencing efforts got under way, Stefánsson flew to Iceland to investigate what kind of support he might have there. After landing, he met with Iceland's minister of health and social security, who administers Iceland's single-payor health care system, and the minister of industry. He traveled through the capital city to speak to scientists and doctors to determine whether enough of them would work with him to make his plan possible. He needed doctors, in particular, for their patients. When he finished, Stefánsson felt he'd found enough backers to put together a company. His biggest cheerleader was his old high school classmate, David Oddsson, who was Iceland's prime minister.

Wanting to be a part of the exciting new project, Gulcher slowly resigned himself to the idea of moving to Iceland. And with his lieutenant largely on board, Stefánsson began fund-raising. Though drug-industry executives had become somewhat skeptical that genetics would lead to cures any time soon, in part because of Haseltine's lobbying, many still believed that the next generation of drugs would be developed based on information about genes.

Stefánsson first looked into the possibility of establishing a non-profit research institute in Iceland, but nobody wanted to fund one of those. So he decided he'd found a biotechnology firm instead. It would be Iceland's first. Aided by his intense charm and obvious intelligence, Stefánsson lobbied American investors. He explained the unprecedented opportunity Iceland presented for finding important human disease genes. He invited venture capitalists to meet him and Gulcher in Iceland to show them what the country had to offer.

Though Icelandic is his native language, Stefánsson speaks eloquent English with a melodious brogue and a Shakespearean tone that alternates from deeply philosophical to ironic to emphatic. His passionate approach proved powerful, raising $12 million in venture capital in less than a year. Three months later, he moved back to Iceland.

ॐ

To a newcomer, landing in Iceland's Keflavík airport is an eerie experience. The surrounding landscape is covered with irregularly shaped black, volcanic rocks that form jagged natural sculptures. A thick spongy moss weaves between the rocks, giving the ground a pale-green tint. Bald mountains, part of a volcanic zone along the Mid-Atlantic Ridge, form a majestic backdrop for this primitive moonscape.

To travel along the desolate road from Keflavík to Reykjavík is to pass through time. The swaths of barrenness give way to colorful town houses, modern museums, trendy shops, tall ships, and people with pale complexions toting pocket phones. But when Stefánsson took this trip in late 1996, he was pushing past this already modernized present and clearing a path to the future.

Some people regarded Stefánsson's return to Iceland like that of a hero from some medieval Icelandic saga. Stefánsson cast a hero's shadow, standing fully six foot five inches tall, his broad shoulders habitually draped with a dark, loose-fitting, unbuttoned jacket. Two decades before, he had been part of the periodic exodus of talented Icelanders leaving for greater opportunities abroad. He was now one of the rare few bringing business back home.

New business was sorely needed. In scattered spots around Iceland, fish still dried on rickety wooden scaffolds, and in Reykjavík and neighboring towns, tall boats dominated the landscape—signs of a society living largely off the sea. Proceeds from fishing made up three-quarters of Icelanders' wages.[2] Stefánsson promised to help broaden that economic base by giving Icelanders a chance to make a living in a radically different way: fishing for genes. He'd lure Icelandic scientists home, helping to halt the drain of talent out of this tiny land.

That process had begun even before Stefánsson arrived. As soon as he'd secured funding back in the United States, he had set up a fax machine in a small office in Iceland. The machine was immediately flooded with letters from Icelanders living overseas or their family members inquiring about jobs that might bring them or their loved ones home.

Stefánsson set up shop on the second floor of a three-story building whose first and third floors housed a Kodak processing plant. It sat in the back of an industrial park on the outskirts of Reykjavík. He filled the rented middle floor with a half dozen automated DNA-decoding machines costing hundreds of thousands of dollars each and five state-of-the-art DNA-copying robots that prepared the genetic material for analysis.

Stefánsson officially opened the doors of his new firm on November 4, 1996. He called it Decode Genetics or, in Icelandic, Íslensk erfðgreining. He already had about fifteen people on board, including Gulcher and some recruits from the faxes, and he set out to hire dozens more.

Among them was a young Icelander named Daníel Óskarsson with a flair for using computers to solve biological problems. Óskarsson was just finishing his master's thesis in Sweden in 1997 when his mother told him about Stefánsson's new venture. Óskarsson was dubious. He wasn't interested in applying to an Icelandic company for work. He figured it had to be a two-man company in a garage.

When Óskarsson visited Decode on a trip home that fall, he changed his mind. He found the place abuzz with some fifty employees engaged in a state-of-the-art research program. And Stefánsson needed someone with Óskarsson's skills to do just the kind of work the expatriate had long dreamed of doing: deciphering the information hidden inside the genetic code. Instead of returning to mainland Europe to find a permanent job, as he'd planned, Óskarsson stayed in his native country and reported for work at Decode.

Stefánsson also courted doctors whose patients he'd need to study the genetic basis of disease. He visited their offices, offering incentives such as credit on scientific papers and a share of any revenue from future corporate deals that made use of the information the doctors supplied.

Some doctors were wary of working with a company, which might put profits above scientific integrity. But Stefánsson persuaded many of them that his team was dedicated to the science, and that good science was critical to their business. Understanding the genetic basis of a disease, he said, was the first step toward designing effective drugs to treat it.

Soon after Decode opened, Stefánsson teamed up with neurologists at Reykjavík Hospital and the National Hospital of Iceland to study patients afflicted with a common neurological condition called essential tremor. This disorder causes uncontrollable shaking of the arms and hands and, sometimes, the head and voice. Patients become progressively less able to perform simple tasks like eating soup with a spoon; most of the afflicted millions, largely people over age sixty-five, are distressed and embarrassed by their quivering body parts.

Among them was Gulcher's mother, who passed a milder form of the trait to her son. Stefánsson's father-in-law, a well-known abstract painter in Iceland, had a tremor so severe he couldn't hold a coffee cup without spilling it. But personal experience with the disease wasn't the only reason for picking tremor as their first project. The ailment was also thought to be simpler than most other common diseases in the number of genetic factors that played a role in it.

Stefánsson felt it would be a good test project, a good way to break in his new employees. No one had yet succeeded in mapping a tremor gene—meaning, in this context, finding its location on one of the twenty-four unique human chromosomes.

For this study, the neurologists assembled family trees the traditional way: by word of mouth. They asked patients if anyone else in their family had tremor, and they examined those family members. They ended up with seventy-five patients and forty-two unaffected relatives in sixteen Icelandic families. The researchers drew sixteen little family trees, representing males with squares and females with circles; and blackened icons indicated those patients with the tremor. Then each consenting individual reported to a clinic to give blood.

Thumb-size test tubes filled with the genetic blueprints of each of the 117 people arrived back at the Decode labs. Each tube of blood was marked with a code disguising the identity of the donor. Inside swam cells harboring identical tangles of DNA—full copies of each person's genetic blueprint. Scientists sucked the DNA from those cells and plopped it in a clear solution, where it floated opaquely, resembling a glob of mucus.

Then, researchers used their robotic DNA-copying machines, which resembled oversize microwaves, to make copies of several hundred small snippets of the DNA, like isolating so many small phrases in a book. Those snippets, copied using breakthrough chemistry known as PCR, the polymerase chain reaction, served as markers that could be used to compare the genetic instruction books of different individuals.

The researchers looked for clusters of identical markers among the DNA of related patients. When such a cluster shows up in patients within a family, the patients are likely to have inherited the same chunk of a chromosome. Where this happens more often than expected by chance—for instance, siblings generally share DNA half the time and cousins one-eighth of the time—scientists consider the area worth investigating as a possible home of a disease gene. In this case, one suspicious spot sat on chromosome 3. The researchers found that this chunk of DNA seemed to consistently follow patients with the tremor in all but two of the families, a result highly unlikely to occur by coincidence.

Decode researchers had found a general location or "locus" for a gene that accounted for the tremor in the vast majority of the Icelandic families they studied. Research groups in England and the United States had already tried and failed to map a tremor gene, probably because tremor had a greater diversity of genetic causes in other populations. The genetic simplicity of Stefánsson's population had paid off. In September 1997, Stefánsson's team unveiled its success in the journal *Nature Genetics*.

"We hope that in Icelanders, one or a few genetic areas are responsible for most cases of a disease," explained Thorgeir Thorgeirsson, one of Decode's scientists. By contrast, he said: "The heterogeneity of other populations means that any one [genetic] location related to a disease may not show up in studies."

The Decode researchers had found a big swath of DNA territory in which the gene for tremor lay hidden, not the gene itself. They planned to analyze the DNA from more distant relatives to gradually narrow down the region until they could pluck out the gene. They knew it would be a tough search, given that everyone had a healthy version of the gene and the damage to the gene in people with tremor might be as subtle as one flipped DNA base pair.

"When we find a locus, it's as if we know the gene is in Scandinavia. Then we will narrow the region to, say, a small country like Iceland," Thorgeirsson reflected. "Next we can start going door-to-door searching for the person—or the mutation responsible for the disease."

By mid-1997, Decode researchers were also hot on the trail of an MS gene, based on DNA samples from some 350 Icelandic MS patients. That summer, Stefánsson promised a reporter: "We're very close now—and we're going to find it."[3] By then, less than a year after its birth, Decode had forged alliances with dozens of physicians for future studies of twenty-five diseases, which it planned to begin once it had enough money.

More or less, the diseases Stefánsson's team had chosen were all complex genetic disorders. Every common disorder in our society—from cancer to depression—fits into this category. Unraveling the roots of such disorders is difficult, because the connection between what's in the DNA code and the appearance of the disease is indirect.

In "simple" genetic diseases, like Huntington's disease or cystic fibrosis, a mutation or change in a single gene produces the disease. Whenever such a mutation appears, so does the disease, and the inheritance pattern is easily recognized in families. By contrast, complex diseases do not result from single genetic mishaps. They are caused by either mutations in several or many genes or a combination of environmental and genetic factors.

For example, genes alone will not cause emphysema. To develop emphysema, a person must also smoke. However, smoking alone doesn't cause emphysema, or at least not in Iceland, because the disease develops in fewer than one-fifth of Icelanders who smoke. The only Icelanders who develop emphysema are those who both smoke and whose DNA bears a certain genetic signature, because emphysema in Iceland runs in families. And it is that kind of elusive genetic signature that Stefánsson's team was trying to read, to unearth the roots of many intractable diseases.

❧

THE TREMOR STUDY was fairly small. And the Decode team knew that word-of-mouth pedigrees would be too limited to map genes for ailments more complex than tremor. For those, they'd need to connect much greater numbers of people in a family tree. They couldn't spare the time and expense required to interview patients about their family medical histories, and they didn't want to miss important relatives due to gaps in patients' knowledge.

The answer was to harness the nation's detailed genealogical records and find the additional family members themselves.

Icelanders have documented their family relationships for centuries. In ancient times, the country was so bitterly poor that it had no statues or cathedrals; instead, it preserved its culture in writing.

Iceland's most prominent surviving legacy is its sagas, tales of a medieval Icelandic society plagued by poverty, battles, and sins, and rescued by heroes. These sagas, while not taken literally as historical fact, nevertheless root modern Icelanders to their ancient heritage. And virtually all of them start with a chapter detailing the genealogy of the main characters.

In 1703, Iceland's rulers conducted the nation's first census, the first complete census of any nation. It yielded an extant familial picture of the population from the preceding half century. The original manuscripts of that census still exist, as do records from the censuses of 1801 and 1910.

In modern times, Icelanders' passion for genealogy is evident in the volumes of genealogical data published every year. Some of these detail the descendants of an Icelandic couple living a century ago while others present pictorial ancestries of present-day Icelandic doctors, lawyers, teachers, taxi drivers, or members of another profession or trade. When a person dies in Iceland, the obituary in the local newspaper includes a detailed family tree that may extend back a century.

This wealth of information was largely scattered among many documents until an Icelandic computer programmer named Friðrik Skúlason began building a genealogy database in late 1988, when his company, Frisk Software International, launched the first genealogy program for Icelanders. The database began as a list of two thousand people who had lived around the time of Iceland's settlement, and were thus more or less the ancestors of all living Icelanders. Skúlason then expanded it to include data from the censuses of 1703, 1801, and 1910; the December 1994 National Registry, a government list of Icelanders who were alive then; and a database of everyone who had died since 1966.

By the time Stefánsson came calling in 1996, Skúlason's database included five hundred thousand Icelanders, and about half of their family connections. Skúlason figured it would take him and his assistant two to three decades to expand the database to include all of the estimated 1.2 million Icelanders who have ever lived and, the hard part, draw nearly all of the connections between them. Stefánsson couldn't wait that long, so he offered Skúlason a deal. Decode would finance a rapid expansion of the genealogy database if the company could use it for its genetics research.

Skúlason jumped at the offer. After a handshake one evening, Decode hired fifteen historians and anthropologists to help build the database, dubbed Íslendingabók, or Book of Icelanders, which would date back to the ninth century A.D. Skúlason oversaw the project and provided access to his extensive genealogical library. In June 1997, they got to work.

Decode researchers soon proved the power of that database in their studies of preeclampsia, a dangerous disorder that causes high blood pressure, seizures, and kidney problems in pregnant women. In mid-1997, Decode began collaborating with obstetricians at the National Hospital of Iceland who had already identified more than one hundred pairs of sisters, both of whom had had preeclampsia during pregnancy. When the Decode researchers examined the sisters' blood, they found the sisters seemed to share a part of chromosome 2. But they didn't have enough people to say for sure whether this was a real site of a disease gene or a red herring.

After the genealogy database became ready for use in mid-1998, they linked sister pairs together to create larger families. And this time, they found more afflicted women in this larger family who shared that piece of the chromosome. In some cases, not only did a patient share the DNA piece with her sister, but also with her cousin and her second cousin, with whom the random chances of sharing DNA are much lower. (They are one-eighth for a cousin and one-thirty-second for a second cousin compared to one-half for a sibling.) With this much stronger evidence, they nailed down the gene's locus.

Despite Decode's success with the genealogy database, Stefánsson had far more audacious plans. He wanted to include more diseases and more people—virtually the entire Icelandic population, he hoped—in his experiments. He also wanted to hasten the pace of his work by eliminating the need to recruit new doctors and patients for each new study. And he sought more complete explanations of disease origins, ones that included environmental culprits as well as genes.

His bold plan: to create a large centralized database with the medical records of almost everyone in the country. The data would come from Icelandic doctors who would systematically enter their patients' health information into local databases that would feed into the centralized one.

Unlike doctor and hospital records, the database would not be used to suggest diagnoses or treatments for individuals, whose identities would be disguised in any case. It would compile statistical information for research purposes—the kind of research that would unearth the origins of disease and catapult the country to worldwide prominence in human genetics.

The company that created the database—assembled the hardware, software, and personnel to make it possible—would provide jobs for hundreds of Icelande[rs. And while lots of] ers would benefit from the data, as aca[   ]cess to it for non-commercial purposes, [   ]tabase, at a cost of $100 million or mor[   ]rmitted to profit from it. Stefánsson wa[   ]

Stefánsson's plan w[   ]recedented. Sweden, Denmark, and N[   ]databases used for both scientific resear[   ]tion. The notion that a private firm rath[   ]run the Icelandic database was new. But [   ]fánsson's scheme was that he planned to marry genetic data to patients' medical records.

The DNA for the database would be collected separately, and only from people who had agreed to give their blood for this purpose. Stefánsson then planned to put ads in the newspaper asking for volunteers, and he was optimistic that the vast majority of the Icelandic population would line up to open their veins. "I bet you anyone's money that more than 200,000 of the 275,000 Icelanders will come and give blood," he later professed. If so, he'd be able to create an extraordinarily powerful tool for decoding the secrets of common, complex diseases.

The tool's potential power came from the combination of DNA and health data. In most common diseases, from heart disease to depression, genes are only half the equation. The environment—smoke, diet, sunlight, emotional stress—also plays a role, and without knowing much about the environmental triggers, Stefánsson felt his theories about the origins of disease were likely to be overly complex or, worse, wrong.

In most instances, the environmental component was likely to be much less obvious than tobacco smoke, Stefánsson reasoned, and he wasn't confident that he or his fellow scientists could guess its identity very often. For instance, he felt scientists could not reliably intuit unexpected connections that have recently come to light like that between infectious agents and heart attack risk. And he didn't want to run after hunches like whether bad parenting leads to schizophrenia. Stefánsson wanted to investigate environmental disease triggers using data and not prejudice.

The data he needed, he figured, could come from a health sector database, since people's health histories typically reflect their most important environmental exposures, though they also planned to have DNA donors fill out questionnaires about those exposures. Combining health care and genetic data, he hoped, would lead to statistical models of the origins of disease that described the interplay between genetic and environmental components. The computer would spit out answers without needing hints about what those answers might be.

Stefánsson assumed the answers would surprise him, just as most of the disease genes that other scientists had found surprised them. Such genes often pointed to a new hypothesis about a disease rather than confirming a previous one. Stefánsson wanted his data to overturn his assumptions. He always welcomed a little chaos.

In presenting a case, Stefánsson comes across as deeply thoughtful, determined, and sincere. Even his ostensible foes say things like: "He's a very beautiful man." As long as nobody is challenging him or irritating him with stupidity, Stefánsson is unusually appealing.

So on the day in 1997 when Stefánsson explained his vision to the Health Ministry, it is not surprising that he endeared the ministers to his idea. The hardest part was convincing them to create a law governing the database, which they didn't want to do and didn't think was necessary. But Stefánsson was loath to commit so much time and money to the project if it weren't protected by special legislation. He spoke to the ministry twice more and met personally with the Health Minister, Ingibjörg Pálmadóttir, to try to persuade her and her fellow ministers to draft the legislation. They eventually relented.

A bill was drafted that gave the minister the authority to grant a single party an exclusive license to construct a centralized database on health care and to market it for twelve years—with the oversight of an independent ethics committee and an appointed operations committee.

❧

STEFÁNSSON THEN FELT welcome in his old country. He was starting a new industry in Iceland, creating not only jobs but also a lot of tax revenue. Ordinary Icelanders seemed genuinely enthusiastic about the potential scientific spoils of the research. And people had begun pouring into the University of Iceland's biology department, because they suddenly saw a possibility of employment in that field.

Of course, a few within the close-knit community of Iceland expressed a tinge of envy at Decode's popularity and early successes. Though the University of Iceland undoubtedly benefited from Decode's presence, university scientists also resented the fact that the company, with its better pay and more advanced equipment, was attracting more bright young researchers and far more of the limelight than they ever could. It didn't help that Stefánsson had, years before, said in an interview that it was the worst school he'd been associated with. The school's staff maintained their distance.

But that icy standoff was positively warm compared to what was in store. Stefánsson was expecting the health care database bill to provoke a debate. But he was not at all prepared for what was to transpire. Stefánsson's bill was about to unleash a wave of intense criticism that would ripple through the small community of Iceland and echo through the rest of the industrialized world.

12

THE DIVORCE

Back in America, Haseltine was fast realizing his own bold vision for producing a cornucopia of gene-based cures. While Decode researchers searched for disease genes one at time and hoped someday to marry those to environmental insults, Haseltine's team rapidly generated vast quantities of genetic material that they hoped could be directly translated into breakthrough medicine.

Since almost every cell in the body eventually ages, wears out, and needs repair, Haseltine envisioned developing a storehouse of molecular glue and replacement parts that could revive every type of cell in the body. He hoped to find natural regenerative substances to grow skin to heal wounds, sprout blood vessels to circumvent blocked ones, and extend nerve cells to limit paralysis or brain damage. He also sought circulating hormones that might tweak the immune system to ameliorate inflammatory diseases such as rheumatoid arthritis, and to fight viral diseases ranging from AIDS to hepatitis. Some natural body proteins might even be used to help halt the growth of cancers.

Belatedly waking up to Haseltine's vision, pharmaceutical industry executives suddenly wanted in on his gene game. They saw genes as the key to the next wave of therapies and diagnostic tests—and a multibillion-dollar pharmaceutical jackpot. They scrambled to claim genetic real estate before HGS and its corporate partner, SmithKline

Beecham, had everything patented. After competitor Incyte Pharma-ceuticals began widely marketing a similar gene database, every major drug company was using the concepts Haseltine and Venter pioneered to start the process of drug discovery.

SmithKline sent its scientists searching through HGS's gene anatomy database for clues to proteins at which they could aim small chemical drugs. In their search for a new obesity drug, SmithKline researchers scoured the HGS database for a gene active only in the brain's hypothalamus, thought to be the seat of appetite. There were a few hundred such genes. However, only a single hypothalamus gene looked like a blueprint for the type of protein likely to be involved in controlling hunger. That was their target. When further research veri-fied the gene's involvement in hunger control, scientists looked for chemicals that blocked it as a way of inhibiting hunger and treating obesity. Company scientists repeated this process hundreds of times, leading to new genes with roles in ailments from cardiovascular disease to migraine headaches.

But soon SmithKline realized it had far more drug targets than it could possibly develop into drugs. So the company decided to sell some of HGS's data. In 1995 and 1996, the HGS family grew to include Takeda Chemical Industries Ltd., Merck KGaA of Germany, the Schering-Plough Corporation, and Synthelabo S.A. (now Sanofi-Synthelabo). Each partner put up tens of millions of dollars for the honor.

By 1996, HGS researchers had spelled out the complete chemical sequence of some three hundred genes and had begun testing their pro-teins for their ability to repair injured parts of the human body. But a potential legal snag stood between Haseltine and his dreams of building HGS into a powerful drug firm. Under its original agreement with SmithKline Beecham, HGS had to give the big drug firm first crack at developing every drug lead its team found (although HGS could always pursue drug leads that SmithKline elected not to pursue). This included not only the targets for small chemical drugs, which were of primary interest to SmithKline because they can be swallowed as pills, but also proteins that Haseltine wanted to develop as drugs themselves.

Haseltine pushed to remove the legal straitjacket, demanding the exclusive rights to develop proteins whose therapeutic value HGS had determined first. SmithKline executives agreed, since they had more than enough to work with and weren't particularly interested in injected protein drugs anyway. But in return, the company demanded that if any of HGS's proteins showed promise in initial tests, Smith-Kline had an option to codevelop and comarket them, starting with large-scale clinical trials.

The HGS team first focused on proteins that bore similarity to proteins already marketed as therapies. Such proteins were likely to come from the same gene family and have similar functions. For example, if a known protein helps boost immunity, other members of its family probably also work as immune system stimulants.

By the end of 1996, HGS was grooming six proteins for possible entry into its nascent drug-development pipeline. They included ones that stimulated skin cell growth to heal hard-to-treat wounds, that protected bone marrow cells from the harmful effects of chemotherapy, and that shielded the nerves that control muscles to limit devastating paralysis such as that caused by Lou Gehrig's disease. So far, the proteins had undergone only limited testing in cell culture and animals, and it was far from certain whether any of them would make it to human testing much less to market. But it was a step toward the dream.

While these molecules underwent further testing, Haseltine took an even bolder step. He instructed his scientists to scour their database for *all* of the body's secreted proteins, not just those resembling known hormones and growth factors. Haseltine's computer wizards wrote a program that recognized a telltale feature of all such proteins—a molecular needle that pierces the exterior of the cell so the protein can exit. In this way, they plucked each secreted protein gene from the database, and HGS biologists cloned and sequenced them before testing them for therapeutic effects.

❧

IN MAY 1995, a *Business Week* story lauded Haseltine and Venter as "The Gene Kings" and displayed both of the men, dressed in matching lab coats and spectacles, on the magazine's cover.[1] By then, however, the gene kings reigned over largely separate kingdoms, as TIGR's work had dramatically diverged from HGS's. Venter had virtually abandoned human EST sequencing in favor of microbial work. In October 1995, a team led by Venter's wife, Claire Fraser, unveiled the 580,070 letters that spell out the genetic instructions of the tiny *Mycoplasma genitalium*, a parasite that lives in human genital and respiratory tracts. With the smallest genome of any known free-living organism, *M. genitalium* promised to yield clues to the minimal set of genes essential for life.

At the same time, TIGR scientists were trying to determine what the seventeen hundred genes they'd found in the bacterium *Haemophilus influenzae* did. They thought the task would be straightforward, but it wasn't. About a quarter of *H. influenzae*'s genes didn't match any genes in any genetic database—they were totally new to science—and the TIGR team could not find a function for another 40 percent of its genes. For the genes for which database matches could be found, scientists could glean possible functions, but they still didn't know how the genes worked together to operate a living cell. So breakthrough though it was, unmasking *H. influenzae*'s genetic blueprint placed them only at the start of the path toward answers, not at the end.

The story was similar with *M. genitalium*'s minuscule set of 470 genes. After scouring databases for clues to the genes' functions, Fraser's team was stunned to find that, again, a quarter of this microbe's genes were totally novel. The theme would play on, with researchers uncovering sizable new sets of genes each time they sequenced a new microbial genome, even among organisms thought to be closely related.

Meanwhile, the Department of Energy developed a keen interest in TIGR's approach for understanding microbes with potential environmental or industrial uses, and it funded Venter's institute to sequence the genome of a tiny creature called *Methanococcus jannaschi*. The organism lived and grew in superheated water deep in the ocean at a steamy 185 degrees Fahrenheit, and required about two hundred atmospheres pressure to survive. Its chemistry, the DOE thought, might be duplicated for use in high-temperature industrial and pharmaceutical processes; its ability to produce methane, a potential fuel, from carbon dioxide and hydrogen also intrigued the agency.

TIGR researchers published the genetic instructions of *M. jan-naschi* in August 1996, the third complete genetic code to be revealed to the world. All three had been done at TIGR, though the first slightly more complex organism—yeast—was not far behind.[2]

&

THE MORE GENOMES the TIGR team sequenced, the more frustrated Venter became by Haseltine's restrictions on the public disclosure of his work, especially now that TIGR was partially supported by government funds. Since both the DOE and the NIH insisted that data generated with their funds be published within six months, the conditions under which Venter accepted a government grant potentially clashed with HGS's rights to TIGR's discoveries.

However, on at least one occasion, Haseltine said that he actually let Venter out of his contractual obligation to wait six months before publishing. In 1996, when Venter asked Haseltine for permission to publish virtually immediately so he could be part of the Human Genome Project's early chromosome sequencing efforts, Haseltine said he gave the okay after clearing that decision with HGS's pharmaceutical partners. Haseltine said that neither he nor his partners felt that human chromosome sequence was commercially valuable anyway.

But to Venter, Haseltine's continued attempts to block the publication of his data seemed to be part of a larger plan to put TIGR out of business, although it's not clear the data dispute had anything to do with Haseltine's prediction that the TIGR-HGS marriage would not last.

By 1997, TIGR had spent enormous amounts of money in legal fees in the battles with HGS. Tensions between Venter and Haseltine ran so high that the two men stopped seeing or speaking to each other. All communications between the gene kings were conducted through lawyers.

Venter wanted out, and Haseltine thought that was just fine. From 1994 on, Haseltine regarded TIGR's research as largely tangential to HGS's mission. SmithKline Beecham also had to be consulted, of course, since they stood to lose TIGR's resources. But that was not a problem, since Haseltine had largely duplicated TIGR's sequencing capacity at HGS.

On June 20, 1997, TIGR and HGS divorced. For Haseltine, TIGR's loss was not a hardship. Haseltine thought of TIGR itself as a liability to his firm, costing it about $10 million per year.

By contrast, Venter took a risky financial blow in the split. Despite the institute's government grants, HGS was still providing 90 percent of TIGR's income. Venter gave up the $38 million HGS still owed TIGR over the next five years under the original ten-year, $85 million contract. But it was a price Venter was willing to pay to extricate himself from a partnership he later likened to a jail sentence. "I gave up $38 million to get away from [Haseltine]," Venter recalled with a grin. "It was the best $38 million I ever spent."

On the same day Venter shook off his shackles, he trumpeted his new freedom by releasing a voluminous store of genetic information over the Internet. It consisted of raw sequence data from eleven microbes, including the nearly complete genetic instructions for the ulcer-causing bacterium *Helicobacter pylori* and a chromosome of the malaria parasite, *Plasmodium falciparum*. Venter also opened TIGR's big EST database after dissolving the agreement that restricted the use of TIGR's EST data by academics.

The release of all this data helped mend Venter's reputation with academics, who had recently become even more committed to rapid release of genome data. At the 1996 Bermuda planning meeting for the Human Genome Project, Robert Waterston and John Sulston had argued vociferously for immediate release of human genetic data. Secret sequences and claims on genetic territory, as had been standard in the past, would not do science any good, they argued. Instead of competing, all researchers now working under the genome project umbrella should work together by sharing their data. A debate ensued, but eventually the genome barons agreed to post their sequence data on international databases every twenty-four hours so the sequencers could not take advantage of it before anyone else could. By the time the gene kings parted ways, that resolution had been made official as the Bermuda Accord.

With Venter's renewed openness, the grant money flowed. In addition to supplying money for small-scale human chromosome sequencing, the NIH also funded TIGR to sequence the genomes of pathogens such as those responsible for tuberculosis and cholera. Microbial genome sequencing projects were invigorating antimicrobial programs in pharmaceutical firms, giving researchers the first new targets in years for revolutionary new antibiotics and vaccines.

TIGR would soon receive more DOE funds to decode the 3.3 million base pairs of the *Deinococcus radiodurans*, a red-hued bacterium that is the most radiation-resistant organism known to humankind. Among the organism's three thousand genes, the TIGR team found the blueprints for an army of enzymes that repair broken DNA. These enzymes may help explain the critter's ability to grow in highly radioactive environments, and thus help researchers engineer organisms that can withstand high levels of radiation at sites of "mixed nuclear waste" and devour the hazardous organic compounds there. To manage the increasing number of projects, Venter had more than doubled TIGR's staff since its inception. It now numbered close to two hundred.

Venter attributed the government's new largesse not so much to his new freedom to release his data but to his distance from HGS. Venter felt researchers had tried to punish TIGR simply because it was associated with Haseltine, whether because of his business objectives or his personality, Venter wasn't sure. "It's hard to separate the two," Venter cheerfully explained one day in his Rockville office. "The main thing I lost by separating from him was that people viewed him as such a flaming a-hole that they liked me by comparison. I actually said to people, you know, I'm now going to start taking the heat for things because I won't have him around to make me look good."

To be sure, there were frustrations. TIGR researchers complained that sequencing the genes of the world's creatures was an endless prospect and they'd like to stop to figure out what all the genes did. There was little funding for such follow-up work. But this was trouble in paradise, a consequence of TIGR's spectacular success in sequencing microbes.

Money worries nagged at Venter. As Venter was acutely aware, government grants could grow and shrink from year to year and offered little long-term security. Venter was looking for ways to build an endowment for TIGR, which then had only a year's worth of operating costs in reserve. And perhaps Venter, a man who was used to taking the heat for things, became just a tad ill at ease as things began going his way.

Over the years, Venter had kept in touch with his friend Michael Hunkapiller, the even-tempered head of the biological instrument firm Applied Biosystems. Hunkapiller had led the effort to produce the first automated DNA sequencer, the one Venter had tested in his lab at NIH in the mid-1980s. Hunkapiller's group had coughed up about $30,000 in free laboratory supplies for Venter and Hamilton Smith's *H. influenzae* project after the government rejected TIGR's request for funding. Every year or so, Venter would fly out to Applied Biosystem's headquarters in Foster City, California, to discuss TIGR's scientific progress and hear about Hunkapiller's latest sequencing gadget or formula. Sometimes they'd joke about when they were going to decode the human genome together.

13

THE MAN FROM BUFFALO

As the Human Genome Project geared up for its first human sequencing, Francis Collins suddenly realized that the lack of attention to exactly what DNA the labs were going to sequence was a big problem. The issue arose in February 1996 at a private retreat in Bermuda, where U.S. and British genome researchers were discussing plans for human sequencing. During the discussion, the question came up: Whose DNA are we sequencing anyway?[1]

Journalists and laypeople had been asking the same thing since the dawn of the Human Genome Project. The official answer was: "No one's." The first human genome to be sequenced was supposed to be a composite of sequences from many sources, including cell lines that have existed in laboratories all over the world for some time. The sequence was thus supposed be a generic sequence representative of humans in general and not of any particular individual.[2]

It sounded good on paper, but it became clear at the Bermuda meeting that no one had thought through the idea. Most of the genome project laboratories trying to piece together the large chunks that were to be sequenced were relying on just one of two DNA libraries: Simon's BACs or de Jong's PACs. From a technical perspective, this was good news. Relying on fewer sources of DNA was expected to make the final sequence much easier to assemble, and both technologies had proven to be excellent.

But Francis Collins was worried. If most of the DNA came from just two libraries—collections of human DNA pieces stuck inside bacteria—researchers would be essentially deciphering the genetic codes of two specific people. This was not the genetic mosaic envisioned in 1990. Had these individuals given their informed consent—that is, consented to it on paper? And, critically, were they anonymous?

Collins and others worried that a lack of anonymity might create a firestorm of controversy if the public questioned the gender or ethnicity of the person representing humanity's genome, even though the choice didn't matter scientifically. The DNA sequence of any two people is extremely similar. Worse, reporters might turn a DNA donor into an unwitting celebrity, exposing him or her to unwanted public attention. The donor's genetic liabilities might even put him or her at risk for denial of health insurance coverage. Since everyone has such liabilities hidden in their DNA, it was not possible to choose a genome without them.

As such scenarios were bandied about, de Jong gulped. Back in 1993, when he'd applied for funds to make his PAC libraries, such matters were far from his mind. It was accepted among his peers that the genome project was about studying our species, not studying any one person. This idea was so ingrained in the scientific culture that de Jong had even indicated on his grant application forms that there would be no human subjects involved in the research. He worked with viruses and bacteria, after all.

None of the committees that reviewed and funded de Jong's grants had asked how his group managed to make human DNA libraries without a human subject. They had been similarly nonchalant about the issue when Simon and other genome librarians requested funds.

As a result, de Jong had simply used the DNA of a person on his staff. Simon and Shizuya had used an anonymous donor's sperm for their libraries, but the NIH didn't fully approve of their procedure for getting informed consent. The DOE's chromosome-specific cosmid libraries were derived from human cell lines that had been developed as early as twenty-five to thirty years ago, so the donors were either unknown or dead, making it difficult to get consent. Even if they could be found, getting permission after the fact was far from ideal since complete anonymity would be impossible and the "volunteers" might feel pressured to cooperate.

Collins felt something had to be done. In May, he called his scientific advisory team and several experts in bioethics together for a confidential meeting. They hashed out the ideal criteria for obtaining consent from those whose DNA would serve as fodder for a significant fraction of the final genome sequence. Over the summer, Collins discussed the proposed criteria with his counterpart at the DOE, Aristides Patrinos.

Collins and Patrinos posted the final guidelines on the Internet in August. Sequencing, they declared, must be done on DNA from anonymous volunteers who have given informed consent for this use of their genetic material. Researchers had to encrypt donated DNA samples, select one at random, and then destroy the encryption scheme so that no one in the future could determine whose DNA had been sequenced. Since no existing DNA library met these criteria, entirely new libraries would have to be made.

This would take time, of course, postponing the day when human genome sequencing could get seriously under way. The lack of forethought about this critical source of DNA thus further postponed the main goal of the Human Genome Project. Many considered this a dull technical matter, no doubt, but preparing an ethically clean, high-quality source of DNA chunks was nonetheless essential to decoding the genome as planned. Indeed, nothing mattered more.

Collins and his advisers finally decided that Simon and Shizuya's BAC technology was the one to use for the main source of human DNA. However, the Caltech team had not sewn up the job of creating the DNA libraries. De Jong also knew how to make BACs, thanks to his early insight that led him to send his laboratory workers to Shizuya's laboratory four years before to learn the technique. Moreover, the persuasive and efficient de Jong had a history of success both in building high-quality libraries and in rapidly getting them out to the community.

De Jong's PACs, in fact, were initially more popular with genome mappers than BACs simply because de Jong and his team reproduced and distributed them so quickly. By the time BACs were declared the vector of choice, de Jong had also started making a BAC library, putting him in the running to contribute the human DNA that would yield the most detailed blueprint of humankind.

In fall 1996, the NIH's genome institute gave de Jong's team a two-year grant to produce two new BAC libraries, one with a man's DNA and one with a woman's. Shizuya and Simon were also funded—by the Department of Energy—on a smaller scale to develop a new BAC library.

Though de Jong was charged with making just two libraries at first, Collins still wanted his mosaic. He reportedly wanted de Jong's and Simon's labs to build numerous BAC libraries—from as many as a dozen donors in all—each of which would provide part of the human DNA sequence. It wasn't obvious how this plan should be implemented since each laboratory could make only one library at a time, a process requiring a couple of years. And even if the librarians could work more quickly, arranging all the library's pieces in order along the chromosomes—the procedure known as mapping—could take several years by itself.

De Jong spent the rest of the fall and the winter of 1997 negotiating with the NIH about how to select DNA donors. With that settled in March, he did something smart: in addition to advertising for donors in the Buffalo newspaper, he approached a local science journalist with a story idea. Buffalo-based researchers were looking for area residents to provide the material for the blueprint for mankind. This was a beautiful chance for Buffalo to enter the national spotlight, de Jong suggested.

The reporter bit. On the last Sunday of March, *The Buffalo News* ran a front-page story about the Human Genome Project and how Buffalo could contribute. The article listed a phone number for people who wanted to donate their blood.

Nearly one hundred people volunteered. A few—including one airline pilot who was on a layover at the Buffalo airport—were so eager to be selected that they disqualified themselves in the process. Unable to reach anyone on Sunday at the listed number, which rang in the genetic counselors' offices, they tracked down de Jong at home. But as a researcher involved in the work, de Jong couldn't have any contact with the people whose DNA he might be cloning. Once somebody called him, he or she was unable to participate.

Monday morning, the genetic counselors' phones rang incessantly. According to the protocol, the first ten female and first ten male callers would be asked to come in and donate their blood. De Jong joked that there might have been some selection for alertness, but the choice of donors was otherwise unbiased. One male and one female donor were selected at random.

Within two months, de Jong's group isolated the DNA from the male donor, stuck pieces of it into the BAC vectors, and coaxed bacteria to copy the pieces. It took another few months to pick the bacterial clones and duplicate the library. That library was called RPCI-11 for Roswell Park Cancer Institute #11. De Jong's team distributed the first copies of the library to the genome project laboratories in fall 1997, just a year after he'd started making them. He beat Simon and Shizuya by months.

By the time Simon and Shizuya's BAC libraries came out, the genome mappers couldn't handle switching to a new library. The genome labs mostly used the Caltech clones to fill gaps left in de Jong's initial library. So instead of a mosaic, the sequence would be derived largely from one person—an anonymous male who presumably once lived in or near Buffalo, New York.[3]

But with a good set of clones finally in hand, the various laboratories funded by the Human Genome Project were at long last ready to start seriously sequencing the human genome. Little did they know that an impending tempest was about to seriously upset their plans.

14

SEEDS OF THUNDER

The day after Venter announced his divorce from HGS in summer 1997, an Applied Biosystems product manager, Stephen Lombardi, called Venter. You ought to think about doing some business with us, he said. He had a number of ideas for a TIGR–Applied Biosystems collaboration.

Venter was busy and didn't give those schemes much thought. But the product manager's call augured a volcanic shift at Applied Biosystems' parent company, one that would shake both Venter's world and that of the genome establishment.

It could hardly have come from a more unlikely source. The parent firm, Perkin-Elmer Corporation, was a tired sixty-year-old business nestled among the small towns of southern Connecticut. It was devoted largely to analytical chemistry, making lab tools for sifting and detecting various substances—checking air and water quality, measuring blood-alcohol levels in urine, and the like. It wasn't glamorous, and it was no longer even very lucrative. That business was contracting.

But Perkin-Elmer had a new CEO, a fiery, blunt-spoken forty-nine-year-old southerner named Tony White. Though he had no formal scientific background, White understood the health care business from his twenty-six years at Baxter International, and he was determined to shake up the stodgy firm.

At the epicenter of that imminent quake was Applied Biosystems, which made DNA-spelling machines used heavily in academic and corporate genetics research. Applied Biosystems' business was rapidly growing, despite the lack of investment from the company's former managers, who were unfamiliar with genetics. Already, more than 90 percent of molecular biologists clambered to buy Applied Biosystems' equipment. White sensed the field had momentum.

Signaling a change in style, White decorated his office in the company's Norwalk, Connecticut, headquarters to feel like a casual den, with wooden furniture and cabinets, and eclectic souvenirs, including a wall collage of pieces from Buddhist temples in Singapore given to him by a friend. And unlike some of his predecessors, White encouraged his staff to speak their minds. He also tuned his ears to his customers and competitors.

And so it was that in 1996, White came to be listening to a speech given by a leader of Incyte Pharmaceuticals, the Palo Alto–based competitor of HGS and a customer for Applied Biosystems' sequencing machines. The speech, according to White, painted a world in which tool makers like Perkin-Elmer were mere commodities—at the bottom of the corporate food chain—and companies like Incyte were molding these raw, relatively primitive commodities into real value that could command a premium price. That value was in data that pharmaceutical companies would buy and use to make diagnostics and drugs. Incyte's chairman said that he planned to force Perkin-Elmer into price competition by helping smaller instrument companies compete more successfully for its business.

White knew Incyte's scenario was partly hot air, contrived to convince investors that Incyte was at the top of the heap. But the speech infuriated him even so. How dare Incyte say we're the commodity when we created the technology that's enabled them to exist? he fumed. Applied Biosystems' sequencing machines were products of great innovation, hardly raw materials anyone with a shovel could dig out of the ground. Applied Biosystems' profits swamped those of Incyte. So who was to say who was at the top of the food chain and who was at the bottom?

Nevertheless, White heard the warning. He resolved to do everything possible to keep his company from being demoted to commodity status. He began casting about for ways to move his business up the corporate food chain. In particular, he pondered becoming a business like Incyte that would produce and deliver biological data.

Mike Hunkapiller, Venter's friend and Applied Biosystems' head, had also considered this from time to time, but his ideas had not gone anywhere under Perkin-Elmer's risk-averse former management. In any case, White's data-delivery dream was bigger than any of those. And now, critically, there was momentum from the top.

But moving into this new area would take a blockbuster idea, something radically different from what others were doing. For a long time, Tony White hadn't a clue what it would be.

The critical insight would come from an innovative life sciences company that White was about to acquire to boost Perkin-Elmer's growth in that area. PerSeptive Biosystems of Framingham, Massachusetts, built tools for scientists to use in studying proteins. It also had a hidden asset: a Lebanon-born bioengineering wizard named Noubar Afeyan. Afeyan founded PerSeptive in 1987 at age twenty-four, just after earning a Ph.D. in biochemical engineering at MIT. By the time Perkin-Elmer decided to buy the firm in a deal worth $400 million, PerSeptive employed seven hundred people and had $100 million in sales.

Afeyan knew he'd soon be a wealthy man. He could walk away from work forever. Anticipating this possibility, White took Afeyan aside and said he'd really like help in transforming his company. Would Afeyan consider sticking around to help him do that? He offered Afeyan the chance to oversee the integration of PerSeptive with Perkin-Elmer. Afeyan liked that idea. PerSeptive was his baby. Afeyan also liked White and his no-nonsense demeanor. He agreed.

White wanted Afeyan's help to make up for his own lack of formal scientific training and scant experience in biotech. But he didn't know the half of it. Before selling PerSeptive, Afeyan had made detailed—but unrealized—plans to parlay the protein-tool business into a data factory. Unbeknownst to either, the two businessmen were on exactly the same wavelength. The bold and feisty, freethinking engineer also had the perfect personality for shaking up White's sleepy firm.

White instructed his new recruit: "Be provocative."

⚮

THE COMMAND WAS fresh in Afeyan's mind as he sat around a U-shaped table among a gathering of Perkin-Elmer board members, scientists, and businessmen in November 1997. They were attending a technology meeting at Applied Biosystems in sunny, windswept Foster City, California. Afeyan was still the CEO of PerSeptive until regulators approved the merger. He had come to describe PerSeptive's technology and to observe. He listened intently to all the presentations and took particular note when Stephen Lombardi stood up to unveil the progress on something called the Manhattan Project.

The Manhattan Project, Lombardi explained, was the code name for an endeavor to build a powerful new DNA sequencer, a machine that could spell out the chemical letters of DNA molecules. Unlike the machines Applied Biosystems had made in the past, which required a person to load them with new DNA samples every few hours, this one included a robot that automatically loaded the DNA to be sequenced. For twenty-four hours at a stretch, it would run itself without need for human input. This would increase the amount of sequence that could be generated per day, reduce labor costs, and relieve any sequencing operation of the burden of finding scarce lab technicians.

At the time, the machine itself was nothing more than parts attached to metal Peg-Boards, but Lombardi had big news. The key to the machine's powers—a system for automatically reading the order of the DNA letters indicated by the machine—now looked feasible. The project was going forward. When Lombardi finished, the room came alive with questions about the machine's capabilities, how much it would cost, and when it would be ready. They planned to introduce it in about a year, at a big sequencing meeting to be held in September 1998.

Afeyan mulled over Lombardi's figures. He carefully considered the projected cost per base pair and the number of base pairs the new machines were expected to sequence per day. Suddenly, he blurted out: "Why don't we do the human genome?"[1]

Some people chuckled, figuring the bizarre utterance was a joke. But Afeyan went on in earnest, addressing Hunkapiller and Lombardi. If the cost of sequencing the genome came down tenfold, from the

$3 billion the U.S. government would spend to $300 million, that was cheap enough that a commercial payback might be possible. Afeyan asked once again, "Why don't we do it?"

"We'd be competing with our customers. That's not what we do for a living," someone said. "We make scientific tools for others. We don't do science ourselves," another objected. But Afeyan could see Hunkapiller was intrigued. White also seemed to be listening. Remembering his role as provocateur, Afeyan pushed a little harder. "Just because that isn't the business we're in doesn't mean it isn't the business we can't be in," he argued.

During the coffee break, people started punching numbers into their calculators. First they calculated exactly how much sequencing they'd have to do. Because DNA is, by necessity, sequenced in small pieces and those pieces are sampled at random from many copies of the genome, some chemical letters are, by chance, read many times before others are read once. The scientists used an equation that told them how many times over a genome must be sequenced, on average, to capture a given fraction of the genome's letters at least once. Covering the genome ten times over, they discovered, would capture 99.99 percent of its letters. So they figured on sequencing about thirty billion DNA bases.

They estimated the number of machines they might devote to the project (maybe two hundred) and the number of DNA bases each machine might sequence per day (maybe three hundred to four hundred thousand). After the break, somebody announced, "You know, we could do this in a couple of years."

White himself voiced perhaps the thorniest question of all: "So what? Is there a business here?"

One researcher thought such a project would wake up the world to the power of genes as a research tool, and, as a result, the company would sell a lot more DNA sequencers. That was nice, White thought, but it wasn't quite enough to justify such an enormous commitment. But he was more excited than he let on. Just maybe this idea could be parlayed into the powerful DNA data firm White dreamed of. Could this be his chance?

The idea of building a large DNA-sequencing factory to sequence the human genome was audacious, to be sure, but Perkin-Elmer did have access to the basic technology, courtesy of Hunkapiller's team, at zero markup. They also had the money to acquire the computer power and topflight researchers needed to carry out the task. They didn't have the biological expertise, though. And while no one mentioned any possible collaborator at the meeting, Lombardi's first instinct was: Craig Venter.

∞

To CREATE A big factory to sequence the human genome had been a dream of Venter's ever since he tested the first automated DNA sequencer at the NIH in 1987. But then such a factory would have been astronomically expensive. Sequencing machines were not the only issue. At that time, researchers lacked advanced technologies for manipulating and copying DNA and the computer software and hardware needed to assemble the sequence. In 1987, it cost as much as $10 to sequence just one chemical letter of DNA. At that price, not even the U.S. government could afford to plunge headlong into deciphering the human genome and its three billion DNA letters. So the pioneers of the government project were waiting for, and encouraging the development of, faster and cheaper technology.

A decade later, the situation had changed considerably. Though no technological revolution had taken place, sequencing machines had gradually increased in speed, and the methods for preparing DNA for sequencing had improved significantly. Government-funded researchers had finally landed upon a reliable and efficient way of copying the large pieces of human DNA they needed for their sequencing strategy to work. By 1997, they had started sequencing human DNA on a small scale, an effort in which TIGR participated, tackling part of human chromosome 16. But progress was slow. Nobody outside Perkin-Elmer knew that all this was about to change, or that Venter's long dormant factory dream was about to become reality.

∞

A day or two after the technology meeting, Venter was sitting in his spacious corner office at TIGR next to one of his giant black poodles when the phone rang. It was Steve Lombardi again, and this time he had a more specific idea. He told Venter to close his door, and he popped the question. Would he be interested in working with Perkin-Elmer to sequence the entire human genome?

Venter laughed. "You've got to be shitting me," he said. "You guys are really going to do this?" The international Human Genome Project had slated $3 billion, fifteen years, and more than a dozen labs for the effort. Venter didn't believe Perkin-Elmer would want to risk the time or the money for such a bold endeavor. Perkin-Elmer was not known for taking big risks. Even if it were, what would be the return on the investment?

Lombardi tried to convince Venter that his bosses were serious. He explained how the new machine would change the equation by drastically reducing the costs and time frame for sequencing. Venter was skeptical. "Yeah, sure," he said, and told Lombardi he had to go. He hung up and thought that would be the end of it.

For New Year's Eve 1997, Venter flew to a resort on Hilton Head Island in South Carolina for Renaissance Weekend, an intellectual smorgasbord and party with U.S. president Bill Clinton and his family. There, among the exclusive invitees—big names in business, the arts, sports, science, politics, and the media—Venter ran into a senior Perkin-Elmer employee named Mark Rogers, who invited him to lunch. Rogers then told him that Perkin-Elmer execs were serious about doing the big sequencing project, and asked what Venter thought. Venter said he was dubious. He hadn't seen the new technology, and in any case, was the company really willing to put up that kind of money? Venter shrugged as he walked away.

One reason Venter didn't believe this plan was serious was that none of Perkin-Elmer's top executives had approached him. The executives were serious about the plan, but they were still wavering about Venter. From a scientific perspective Venter was the top choice, given his group's superb sequencing skill. But from a business standpoint he wasn't. An alliance with a nonprofit organization like TIGR was problematic. A large influx of corporate financing would jeopardize TIGR's nonprofit status.

Charles DeLisi, now teaching at Boston University, launched the Human Genome Project at the Department of Energy in the mid-1980s. *Courtesy of Boston University Photo Services*

Nobel laureate James Watson, who with Francis Crick determined the basic structure of DNA, took over as the first head of the NIH's portion of the Human Genome Project in fall 1988. *Courtesy of NHGRI*

Nobel laureate Hamilton Smith spearheaded the first successful effort to decipher the genetic instructions of a living thing. *Courtesy of Applera Corporation*

Cells translate genes into proteins through an intermediary called messenger RNA (mRNA). *Reprinted/adapted by permission from Neil R. Carlson,* Physiology of Behavior, *7th ed.,* © 2001 *by Allyn & Bacon.*

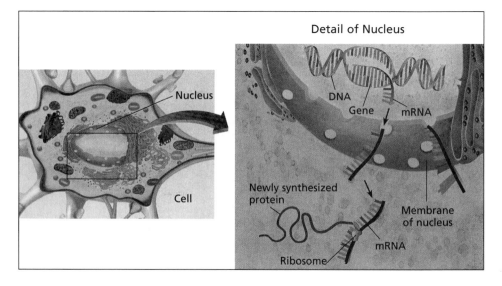

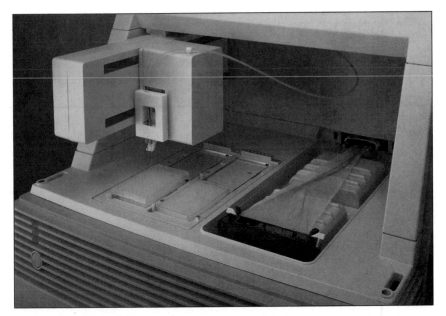

This robotic DNA decoding machine, the PRISM 3700, was used both by privately and publicly funded scientists to decode the complete human genetic instructions. *Courtesy of Applied Biosystems*

Researchers in Buffalo, New York, advertised in the local newspaper for volunteers to donate the DNA that Human Genome Project researchers would use to unravel the human genetic blueprint. *Courtesy of Pieter de Jong*

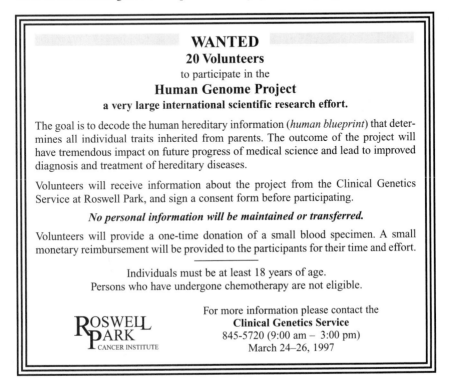

WANTED
20 Volunteers
to participate in the
Human Genome Project
a very large international scientific research effort.

The goal is to decode the human hereditary information (*human blueprint*) that determines all individual traits inherited from parents. The outcome of the project will have tremendous impact on future progress of medical science and lead to improved diagnosis and treatment of hereditary diseases.

Volunteers will receive information about the project from the Clinical Genetics Service at Roswell Park, and sign a consent form before participating.

No personal information will be maintained or transferred.

Volunteers will provide a one-time donation of a small blood specimen. A small monetary reimbursement will be provided to the participants for their time and effort.

Individuals must be at least 18 years of age.
Persons who have undergone chemotherapy are not eligible.

ROSWELL PARK
CANCER INSTITUTE

For more information please contact the
Clinical Genetics Service
845-5720 (9:00 am – 3:00 pm)
March 24–26, 1997

Members of the new U.S.-British gene team met in Houston in February 1999 to garner their forces to complete a working draft of the human genome by spring 2000, ahead of the competition. Attendees were (*back row, left to right*) Michael Morgan (Wellcome Trust), Elbert Branscomb (JGI/DOE), Richard Gibbs (BCM), George Weinstock (BCM), Francis Collins, Eric Lander (MIT), Mark Guyer (NHGRI), Greg Schuler (NCBI), (*front row, left to right*) Elke Jordan (NHGRI), Lauren Linton (MIT), John Sulston (Sanger Centre), Jane Peterson (NHGRI), and Adam Felsenfeld (NHGRI). *Courtesy of Louis Adame, Baylor College of Medicine*

William Haseltine, head of the biotech firm Human Genome Sciences, set out from the start to develop a revolutionary new class of gene-based medicines. *Courtesy of Human Genome Sciences, Inc.*

The Billion Base Bash: Robert Waterston and his genome team at Washington University in St. Louis celebrate the completion of one-third of the human genome, a milestone Human Genome Project scientists reached in November 1999. *Courtesy of Washington University School of Medicine in St. Louis*

Gene maverick Craig Venter—shown in 1995 at TIGR—went on to challenge government scientists to a race to decode the human genome. *Courtesy of TIGR*

In Celera's NASA-style command-and-control center, employees monitored the output from the sequencing factory upstairs, where hundreds of machines spelled out the 3 billion chemical letters of the human genome. *Courtesy of Applera Corporation*

Top Celera researchers Hamilton Smith and Mark Adams in the company's sequencing facility. *Courtesy of Applera Corporation*

In a heroic solo effort, computer wizard James Kent assembled the human genome sequence for the Human Genome Project in his garage in about a month.
Carter Dow

LEFT: Kari Stefansson's bold plans to unravel the genetic secrets of the people of Iceland have met with admiration, awe, and fury.
Northphoto.com

Renowned biologist Claire Fraser now leads the scientific institute TIGR, which her husband, Craig Venter, founded in 1992.
Courtesy of TIGR

President Bill Clinton with Francis Collins (*left*) and Craig Venter at the White House on June 26, 2000, to celebrate the preliminary drafts of what Clinton called "the most important, most wondrous map ever produced by humankind." *Courtesy of NHGRI*

Celera's Craig Venter (*left*) and the NIH's Francis Collins both clasp hands with Aristides Patrinos of the DOE at a press conference at the Capitol Hilton celebrating the completion of the rough draft of the human genome on June 26, 2000. *Courtesy of NHGRI*

Nevertheless, White and Hunkapiller eventually decided to talk to Venter in the hope that they could find a workable financial arrangement. Hunkapiller had known Venter for a decade, from the days when he'd given him his first sequencing machine. Venter had had a history of doing things one step beyond what most people thought possible. Hunkapiller figured this project was right up Venter's alley.

When Hunkapiller called Venter in late January, Venter knew something serious was up. Hunkapiller provided the specifications for the new sequencer over the phone, and Venter agreed to come see the technology. A week later, Venter flew out to California with his top lieutenant at TIGR, Mark Adams, and one other top aide.

On the plane, Venter and Adams discussed the project seriously for the first time. Venter calculated the number of pieces of DNA they'd need to sequence, and he showed his figures to Adams. The job looked possible, Venter thought. Adams hesitated. The numbers said they'd have to sequence some two million pieces of DNA a day. Adams, who'd been in charge of the details of all previous sequencing projects, realized that would require shuffling and storing thousands of little plastic plates filled with DNA. No way. That would require a facility as big as a small airport.

In Foster City, Hunkapiller and Lombardi sequestered Venter and Adams inside a large conference room, where they presented the latest data on the top-secret DNA-sequencing device. Then they led Adams and Venter into a locked, windowless room of another building to take a look at the actual machine. It wasn't pretty. Without a case, the machine's metallic guts were exposed, but the engineers promised it would work. Venter and Adams were impressed by reports of the machine's automation, and its potential for greatly reducing the manpower necessary for DNA sequencing.

Back in the conference room, they started calculating the feasibility of the plan. They soon discovered Venter had made a calculation error on the plane. His estimates had been ten times too large. They'd need to read only two hundred thousand DNA fragments per day, not two million. Suddenly, it all seemed very possible. No matter that TIGR's sequencers then read at a rate of just twenty-five thousand fragments per *week*.

The scientists took turns plotting out various scenarios. If they lined up some two hundred of Perkin-Elmer's fancy new machines, with each one spelling out a thousand DNA fragments (half a million DNA base pairs) per day, they could sequence the entire human genome ten times over—thirty billion base pairs—in just three hundred days, or less than a year. Of course, they'd need additional time to get up to speed, and the machines would not always work properly. But even so, they figured they could do the job in less than three years!

Adams was amazed. That was quite fast and the costs—heavily discounted since Perkin-Elmer would supply the most expensive materials—seemed reasonable, too. He was still worried about whether they'd be able to feed pieces of DNA to the machines as fast as the machines could sequence them. But an Applied Biosystems' scientist promised that his group was perfecting other technologies that would greatly speed up those steps in the process.

Within four hours, the TIGR team had completely shed the doubts they'd had back in Maryland. Venter and Adams were blown away by the new technology.

They were not even fazed by the risky sequencing strategy they felt they needed to pursue. In Foster City, Adams had rejected the suggestion that they use the conservative two-step mapping approach sanctioned by the genome project. Adams and his team had used that approach to sequence a small portion of human chromosome 16. Forty TIGR researchers had spent two years sequencing about one-eighth of the chromosome, one of the smallest human chromosomes. It was so painful that Adams couldn't imagine doing it on a massive scale.

Not only was the method slow, but it was impossible to automate. The preliminary steps of making a library of big chunks and then ordering the chunks had to be done by hand. Why bother, when you could just rip the entire genome to shreds, sequence the shreds, and use a computer to piece it back together again? This one-step whole-genome shotgun method had worked for *Haemophilus influenzae* and for other microbial genomes. Why not use it for the human genome, too?

Adams wasn't stupid. He knew that just because computers could piece together the two million chemical letters contained in the genetic instructions of a single-celled creature didn't mean they could do the same for the three billion letters of code in a human being. For the human they'd have to shred tens of millions of identical copies of a person's complete DNA, creating more than seventy million small DNA fragments that a computer would have to reassemble. The sheer number of pieces wasn't the only issue. Unlike bacterial genomes, the human genome contains lots of nearly identical repeat sequences that make it easy to accidentally patch together two pieces that look like they go together but don't. One might never be able to put the human genome back together again.

But in early 1998, Adams and Venter figured modern supercomputers were up to the task. With enough computer power, they said, whole-genome shotgunning would be the better approach—faster and far more amenable to the industrial process they had in mind.

Hunkapiller was stunned by the idea of applying this risky strategy to the human genome, but he didn't object. He nervously decided that he'd trust the judgment of his prospective partners for the moment.

∞

HAVING ALL BUT snared the scientific team they needed by early February, Perkin-Elmer execs now had to figure out how to pay for the factory. Venter was still not completely convinced that White and Hunkapiller would really cough up the money. The venue for the project hadn't even been determined. Could it somehow still be done at TIGR? At Perkin-Elmer? Would a new company or institute have to be formed?

Back in Maryland, Adams and Venter began making concrete plans. A new building on the TIGR campus was already in the works, and they asked the architects to change the design to accommodate a new facility for the human-sequencing project. The architects were confused by the request, and the new design wasn't far along when White came out for a visit.

White wanted to test Venter's enthusiasm for the project and his willingness to make it commercial. Perkin-Elmer was not interested in funding TIGR to do a science project, and White wanted to avoid repeating the HGS-TIGR fiasco. Commercial or no, how could Venter refuse? This had been his dream for a decade. Venter told White: "I have to be a part of this. This can't happen without me."

Venter had his own financial goals as well as scientific ones. His biggest concern was the longevity of his research institute. He knew that money was imperative for doing innovative science, that he would never have been able to develop any of his novel ideas without the significant resources required to do the experiments. Venter had been in talks with businesspeople in hopes of generating more revenues for TIGR. He told White that he hoped to create an endowment for the institute of at least $200 million. White agreed to give Venter 10 percent of the equity stake in the new venture. Venter split it fifty-fifty between himself and TIGR.

White also promised to provide $300 million in start-up money to found the new venture. Venter, who saw himself as a scientist not a businessman, still largely viewed the $300 million as a research grant. He hadn't yet completely embraced the notion that this was, first and foremost, a moneymaking venture.

White and Venter came closer to an understanding during a vacation in the Virgin Islands in March. White was going to St. Thomas with his family while Venter and his wife were sailing their eighty-two-foot-long yacht, *The Sorcerer*, on a Caribbean tour. Upon realizing their vacation plans converged, they decided to meet.

Their rendezvous point was the tip of Virgin Gorda, at the exclusive Bitter End Yacht Club, where guests are ferried to the beach, shops, or restaurants in canopied golf carts. The two men met for dinner with their families, and they hit it off. After dinner, White and Venter walked around and talked some more.

Ever mindful of his stressful skirmishes with Human Genome Sciences, Venter brought up a serious precondition of his involvement in the project: They had to release the human genome to the public.

"That's fine," White responded, "but now you've got to come up with a business plan where you can take $300 million, sequence the genome, give it away for free—*and make money*." Venter vowed to give it some thought.

When they returned from vacation, White and Venter, along with Afeyan, Dennis Winger (Perkin-Elmer's CFO), and Peter Barrett (its business development officer), began hammering out a business plan. They would glean any patentable information from the genome before giving away the raw sequence for free. They would sell different packages of proprietary information derived from the genome at different prices to different clients, depending on their commercial value.

Meanwhile, Venter and Adams had let a few key TIGR employees in on their secret, including Hamilton Smith and computer wizard Granger Sutton, who'd devised the ingenious software necessary to assemble the *H. influenzae* genome. Adams had a hard time telling top sequencing researcher Rhonda Brandon. He didn't know how to begin. The idea sounded so ludicrous.

When asked what he thought of the idea of assembling the human genome from little pieces, the laconic Sutton said he'd have to think about it. He went back to his boss and told him it was doable, given enough computers and manpower. That was his out.

Venter wowed Smith with stories about the incredible new sequencers that could each spell out up to half a million base pairs per day, which was equal to about half of TIGR's daily output in 1994. Venter said the machines would crank out sequence for twenty-four hours at a stretch without human intervention—and that he planned to use them to shotgun the human genome.

Smith hadn't seen the new machines and so didn't know what to think. Sure, he agreed, the shotgun method was the most efficient strategy. But privately he wondered whether it would work. Even so, he was impressed once again by Venter's courage. "A lot of people would have been scared off," Smith recalled. "But as is so typical of Craig, he jumped at the chance."

Researchers, including Smith and Venter, had written papers espousing variations of the two-step mapping approach for sequencing the human genome. But almost nobody dared advocate in print whole-genome shotgun sequencing as a way of deciphering this enormous molecular treatise.

There was one exception, however. Like many others, Smith considered this article only long enough to read the title before throwing it in the trash. Now Venter asked him to reconsider it. The much-maligned paper had suddenly become a hopeful manifesto to be pored over for tidbits of wisdom. It was, in short, a place to start.

15

THE MALIGNED MANIFESTO

Like most other research in the area, the article Smith had read about shotgunning the human genome had its origins in government-funded genome research. It grew out of the genius, and impatience, of a geneticist named James Weber of the Marshfield Clinic in Marshfield, Wisconsin. Weber had spent his entire career mapping genes linked to diseases from Alzheimer's to colon cancer. He felt that every day of delay cost society potential advances in medical research and health care, advances that could save lives.

In mid-1995, as the government was gearing up for its initial human genome sequencing, Weber was already restless. While his colleagues fretted about whether the human genome sequence would be complete and accurate, Weber—like John Sulston and Robert Waterston—was anxious to get the job done.

Even before Adams and Venter, Weber had a hunch that sequencing methods and computer technology had improved enough since 1986 to make shotgunning the human genome feasible. Somebody would have to create the software to piece the little DNA slivers back together, but that seemed possible. Weber also realized early that eliminating the time-consuming steps of preparing and mapping bigger pieces of human DNA would greatly speed up the process of sequencing the human genome.

Weber recruited his coworkers in Marshfield to help draft a long application to the NIH for a grant to use the method on the human genome. The NIH was soliciting applications for pilot human genome-sequencing projects, and Weber didn't want to miss his chance to be a part of that effort. In the application, which Weber submitted on August 3, 1995, he asked for $12 million over three years, after which his group would have sequenced 1.88 billion scattered base pairs of human DNA. By 1999, if all went well, they planned to sequence enough bases to cover 90 percent of the genome.

Twenty-four days after submitting the application, Weber got a taste of his chances. He floated his shotgun idea at a genome project planning meeting near Dulles Airport, where about two dozen genome barons, including Waterston, Olson, and Collins, discussed whether or not it was the right time to start serious sequencing. The participants responded coolly to Weber's suggestion. One of the barons dubbed the idea "fanciful."

But Weber was undeterred. He set out to recruit the help he needed to prove the barons wrong. Two months before, a colleague of his at the National Center for Biotechnology Information in Bethesda, Maryland, had recommended Eugene Myers at the University of Arizona in Tucson as a potential collaborator on the project. Like Granger Sutton, Myers was part of a then relatively rare breed of computer scientists who work on biological problems. And he was one of the best. Weber's friend had said that Myers had the expertise to work out the difficult computational details of stitching together tens of millions of random overlapping DNA sequences. He could determine if the task was indeed feasible from a mathematical point of view.

Weber phoned Myers and asked for his help.

Normally a talkative man, Myers was suddenly at a loss for words. As it turned out, Myers had been thinking about the mathematics of the whole-genome shotgun method. Several months before, the Wisconsin geneticist Frederick Blattner had asked him to help prepare a proposal to shotgun the genome of the bacterium *Escherichia coli*. In doing so, Myers had already disproved one of the arguments biologists often made against the method.

Biologists had long assumed that shotgunning would require far more DNA sequencing than the two-step mapping approach. But Myers had realized this assumption was wrong. With both methods, DNA pieces from many copies of the genome are sampled randomly. So by chance, some chemical letters are read many times before others are read once. Myers had taken a second look at the equation that tells biologists how many times a genome must be sequenced, on average, to nab almost all the bases.

According to the equation, the size of the piece of DNA in question doesn't matter; the percentage coverage for a given level of sequencing remained the same. For a DNA piece of any size, tenfold average coverage—in which some letters sequenced twenty or more times and others just once—secures 99.99 percent of the genome. Threefold coverage captures 95 percent of the bases, and onefold, 63 percent. The two-step strategy in which the human genome is first broken up into chunks required the same amount of sequencing as the shotgun method of shearing the whole thing into tiny pieces in a single step.

Individually, the chunks would have fewer gaps in their sequences than would the entire, shotgunned human genome. But taking into account the size of the chunks relative to that of the genome, the total amount of missing information would be exactly the same. Thus, the fear that shotgunning really big pieces of DNA would yield a product full of holes was unfounded.

However, Myers still worried about the assembly problem. He imagined somebody dumping seventy million pieces in front of him and saying: "Okay, smart ass, what are you going to do with these?" He told Weber he'd get back to him.

First, Myers calculated how long it might take a computer to do the assembly, comparing thirty billion letters of data against one another, a problem bigger than had ever been tackled. The answer: seventeen computer years. Not bad, he thought. With one hundred computers running simultaneously, the job might be done in nine months.

Then he considered what other sources of information he might use to help put the pieces together. Aside from the overlapping sequences, he might pair up strings of letters from both ends of each little strip of DNA, which can be sequenced from either end. If the sequencing machines read about five hundred base pairs from both ends of each little piece of sequence, the computer would know that those two sets of five hundred base-pair sequences were linked, separated by a fixed amount. That, Myers thought, might give him the handle he needed.

To gain proof that the feat was possible, Myers simulated the problem on a computer. He created a synthetic genome that contained both repeat sequences and unique sequences just like the human genome. He asked the computer how many of the unique sequences it could link together using overlapping sequences, the pair information, and some longer-range linking information from other sources. The answer depended on how much sequencing had been done. If the genome had been sequenced six times over on average, yielding 99.75 percent of the genome, the computer could assemble all but .05 percent of the unique sequences. Myers was surprised. This is really great, he thought.

The computer genome didn't account for all the potential land mines in the human genome. It assumed the repeated segments were short, for example, and fairly evenly distributed, which might not be the case. But Myers thought it was a fairly good test of the approach. He sent Weber his results and agreed to collaborate.

Myers and Weber began working out more of the details in preparation for a paper they planned to write together. Both were convinced that they had landed upon a better strategy than the one that genome project participants were pursuing. They hoped to make the world of geneticists see the light.

Their first chance came in late February 1996. Weber was invited to the international sequencing strategy meeting in Bermuda (the same meeting at which the ethical snafu over the DNA libraries surfaced) to describe their proposed shotgun strategy in detail. Weber asked the meeting's organizing committee to invite Myers, too. Myers could help explain the math behind their proposal. But the committee said no, a decision that angered Myers.

Nevertheless, Weber was given a generous forty minutes to tout his idea. All the key players in the field heard the presentation, including Craig Venter, who was there because of TIGR's early human chromosome sequencing. Weber used all the time he had. To be a major player in the sequencing of the human genome was his dream. But his colleagues thought his proposal was too risky. Someone suggested it be tried on a smaller scale, like a single chromosome, before attempting it with the entire human genome. Weber felt there wasn't time.

Weber was drowned out not only by voices of caution but also by the passionate cries of those heavily invested in strategies inconsistent with his. Collins made it clear that the genome project was not going to abandon its more traditional mapping approach, not even for an experiment. After the meeting, Weber felt beat up and totally defeated.

When April rolled around, funds were doled out to a half dozen labs for pilot human-sequencing projects. But Weber's team was not one of them. Though not surprised, Weber was terribly disappointed. A great ship shepherding medical progress had launched, and they had missed it.

Myers was furious. Yes, the whole-genome shotgun method was high risk. He himself had warned listeners in talks that nobody could know until the end whether they'd successfully put the genome together—after spending years, and many millions of dollars, collecting data. The approach did not offer immediate gratification. But it was not a ridiculous idea. It at least warranted an exploratory pilot study. If the government had planned on spending $3 billion on the Human Genome Project, he thought to himself, what would it have hurt to spend $10 or $20 million to see if this method would save them some money?

Meanwhile, Myers and Weber finished their paper and submitted it to prominent journals. *Science* rejected it. So did *Nature Genetics*. Reviewers for both journals said their proposal was unworkable. Myers suspected the objections were motivated by political considerations as much as scientific ones, as they were, if anything, based on gut intuition rather than a thoughtful investigation of the problem.

Finally, Weber and Myers revised the article and sent it off to the journal *Genome Research* nearly a year later. After several rounds of editing, the journal accepted the manuscript subject to the stipulation that it be immediately followed by a rebuttal saying the method wouldn't work. On May 25, 1997, Myers wrote in his journal: "A first for me: A paper so controversial it was coupled with a paper presenting an opposing point of view."

Like Hamilton Smith, most people who saw the paper already had a bias against the approach and so didn't take the paper seriously. But nine months later, when Smith decided actually to read the article after hearing about Venter's radical plans, its reasoning started to sink in. Smith thought to himself: Well, maybe this is possible.

16

THE FINAL PLANS

Even if the shotgun-sequencing strategy would work, Venter had trouble convincing others of the wisdom of teaming up with Perkin-Elmer. Even his wife didn't like the idea. She feared a reprise of the HGS saga and renewed conflict over how much of, or whether, the data they collected would be released publicly. She repeatedly tried to talk her husband out of becoming involved. Craig protested that Perkin-Elmer was different than HGS. Fraser sighed. "Haven't you learned anything?"

Similar concerns popped up when Venter tried to interest other academics in collaborating with or joining the new venture. To them, human sequencing for a profit was anathema because that suggested this genetic Holy Grail would be locked away, with keys given only to those who could afford to pay for them. Most academics would be shut out, they feared, and thus would not be able to use the information to make medical advances.

Early on, Venter tried to convince David Cox, a prominent geneticist from Stanford University, to join his new team. Cox told him that the project sounded great, but that he wasn't interested because he knew the data would not be made public. Venter tried to convince him otherwise, but Cox didn't believe him. It seemed implausible that a company could give its major product away. "Come to the

Perkin-Elmer board meeting," Venter urged. Cox agreed, since he wanted to do what he could to make Venter's promise of public data come true.

On Friday, March 27, 1998, Cox and Venter met in Palm Beach, Florida, for the off-site board meeting, where Venter and White would unveil the sequencing plans to the company's board for the first time. At dinner the night before the board meeting, Cox spoke. He explained why it was important to give the public unrestricted access to the human genome sequence. The greatest value from that sequence comes not from the sequence itself, which is like the clay sculptors use, but from what you did with it. If somebody controls all the clay, he said, there would be no statues, and nobody wins from that.

The next morning, White detailed the still-emerging business plan. The idea, he explained, was to create an information service business, a Web-based DNA information database to which customers could subscribe or pay per view. The raw human DNA sequence they collected would be made public and would serve as the springboard for their Web-based database. They'd make their money, he explained, by interpreting and analyzing that publicly available sequence, along with some proprietary data, for users who couldn't do that themselves. The idea was modeled after Bloomberg's financial information service, which analyzed public financial data for popular consumption. One corporate board member had another analogy. "Let me get this straight," he said. "We can't sell the Bible but we can make people pay to go to church?"

The scientists smiled and nodded. That pretty much summed it up. Cox left the meeting optimistic.

Exactly what DNA data the company was going to sell was still a bit of a mystery. However, White felt comfortable for the moment with the broad outlines of the compromise he and Venter had reached. He felt that if they could actually sequence the entire human genome before the public project could, they would find a way to make money.

However, White still felt shaky from a technical standpoint. Sure, Hunkapiller and Venter said it was possible, and he trusted them—to a point. But he wasn't about to risk his career and reputation on the opinions of two people with a vested interest. In late March, White asked two experts—board member and Yale geneticist Carolyn Slayman and renowned molecular biologist Arnold Levine, then at Princeton University—to convene a panel of experts to hear Hunkapiller, Adams, and Venter present their technical strategy. Unlike the panels that reviewed Weber and Myers's proposal two years back, this panel was independent of genome project scientists who were committed to another method. White wanted this panel to answer a deceptively simple question: Will it work?

A week or so later, Levine reached White in his car on his way to the airport. Levine was unequivocal. The bottom line: "We've done the math sixteen different ways. It will work." He pointed out the biggest risks, but added: "We think they're all surmountable." They were going to build a factory that would gather genomic information at a pace and price that nobody before imagined possible.

On April 15, scientists and businessmen from Perkin-Elmer convened at TIGR to determine how to put the plan into action. Over two days, they discussed the space and personnel they would need, the number of machines and computer power required, their budget for the first few years, and, roughly, the return on their investment. They wrapped their conclusions into a formal proposal for the board.

Meanwhile, Venter and White had to consider the potential political fallout. Companies such as Incyte that also sell genetic information might feel endangered. But White did not think his new enterprise would be a direct threat to Incyte. He was not planning to sell pieces of genes, as Incyte did. Incyte also had only one kind of customer: the rich pharmaceutical company. White planned to sell a greater variety of genetic data to a variety of customers, including some as modestly financed as high school biology teachers. There was so much data to sell, so much biological information unknown, that White figured his venture's presence should hardly be felt in Incyte's corner of the market.

The leaders of the Human Genome Project might be even more anxious about the new commercial player in the field. White didn't think his venture should immediately supplant the federal project, given all the risks and uncertainties the corporate effort faced. Still, he knew its presence might make life more difficult for those entrusted with the job of convincing the U.S. Congress to fund the genome program. And the public project's leaders, who until then had been assured their places in history, might now have to share the accolades.

Even so, White naïvely felt any political problems would be brief and surmountable. The government program aimed only to provide a sequence that others would interpret as they wished. If the new Perkin-Elmer enterprise wanted to do some of that interpreting and sell the results to a variety of customers at affordable prices, why couldn't they? He also figured genome project leaders would jump at the chance to work with his new venture and thereby complete human sequencing much sooner than their target date of 2005. Venter felt he could produce a workable draft of the human genome in less than half the seven years the government had allotted for finishing its human sequencing.

Venter also had reason to think the response to the announcement would be reasonably positive. Before the board meeting, he'd called Aristides Patrinos, who ran the Department of Energy's portion of the Human Genome Project, and revealed his secret, asking for advice. Venter trusted Patrinos, a longtime friend who'd been very supportive of his efforts in the past. Venter's computationally daring method of sequencing microbial genomes had resonated with Patrinos, who'd been trained as an engineer. The DOE was even funding much of that work. Patrinos greatly admired his friend's willingness to take risks and to challenge established dogma, which Patrinos felt represented the best in American science.

Patrinos was enthusiastic about the new venture. What a fantastic way to shake up the system, Patrinos thought, and to create competitive pressure that would spur the field along. Though he knew helping Venter succeed might speed his own project's demise, Patrinos plied Venter with advice on putting his venture together and interfacing with the public genome program. He also kept the plan secret.

On Friday, May 8, White called a special board meeting to approve the project. The board gave its okay. The decision would be announced to the public that weekend.

17

CONFRONTING THE GOVERNMENT

But before a public announcement was made, Venter and his Perkin-Elmer supporters felt they needed to break the news to Francis Collins and his boss, NIH director Harold Varmus. Venter quickly called both men, saying he needed to meet with them the same day.

Venter and Hunkapiller first spoke to Varmus in his office in Bethesda and then drove to Dulles International Airport outside of Washington, D.C., to meet with a harried Collins, who was on his way out of town. Venter knew the meeting with Collins could be difficult. Collins's program would be directly affected, and Collins was far more competitive with Venter than was Patrinos, his DOE counterpart.

Venter and Hunkapiller found Collins and two of his staff members at the United Airlines Red Carpet Club. Collins listened in disbelief as they outlined their plans. It wasn't clear to him how much of this talk was real or realistic, and the scientific strategy seemed highly dubious to Collins. Even so, it was terrible news.

Until then, the genome project had been going well. The publicly funded centers were just wrapping up their human-sequencing test projects and getting ready to expand them. The coming fall, the NIH planned to increase funding for sequencing to $70 million per year from the $40 million average yearly sum doled out during the earlier phase. Collins was deeply troubled by the idea of somebody else attempting to duplicate his teams' work.

Hunkapiller and Venter tried to reassure Collins and his team that the new company was not a threat to the government program, but would constitute an additional, collaborative effort. It was not reassuring, however, when Venter piped up with his idea of what collaboration might mean.

Since his company was going to do the human genome, Venter asked Collins: "Why don't you do the mouse?" The idea seemed logical to Venter. The public's genome project had sequenced only 3 percent of the human genome, and some of the sequencing centers were proceeding much more slowly than expected.[1] Venter saw no clear plan or technology for speeding up the public effort.

But Collins was appalled by Venter's logic. Genome project researchers had spent years laying the groundwork for human sequencing, and, in his view, progress hadn't been at all slow. The current frustrations were natural growing pains, he felt. There was no justification for Venter's group essentially taking over the government's flagship endeavor. The mouse, indeed.

After the meeting, Collins and Hunkapiller found themselves on the same flight to California. Collins took the opportunity to grill Hunkapiller on the new, faster sequencing machines that formed the technological basis for the new firm. Would Perkin-Elmer also make those machines, the so-called PRISM 3700s, available to government-funded researchers?

Hunkapiller assured him that they would. The new firm would get the first instruments, but the technology would later be available to anyone who wanted it. Both men tried their best to be reasonable. Neither was looking for a fight.

But they got one splashed on the front page of *The New York Times*. Perkin-Elmer had given *Times* reporter Nicholas Wade a newspaper "exclusive" to break the story of its announcement. It appeared the next day, Saturday, on an early Web release of the Sunday, May 10, 1998, *New York Times*. None of the leaders of either sequencing project was quoted as saying anything inflammatory. Still, the article emphasized the potential peril to the federal Human Genome Project, saying that the new venture would "outstrip" it, "and to some extent make [it] redundant." It went on to suggest the possibility that Congress might wonder why it should continue to finance the federal effort if the new venture were so much speedier: it was to complete the sequence in a mere three years, by 2001, rather than the genome project's then goal of 2005. It was also to be done ten times more cheaply, at a cost of $200 to $300 million instead of $3 billion.

The article was like a hand grenade thrown in the middle of the NIH campus. In emphasizing the superiority—and threat—to the government program, the article immediately put Collins and Varmus on the defensive. It also dealt a blow to Waterston, Sulston, and other sequencing gurus who had cast their lot with the public program.

On Monday, Venter, Hunkapiller, and White, along with Collins, Varmus, Patrinos, and Patrinos's boss at the Department of Energy, Martha Krebs, gathered at the NIH for a hastily arranged press conference. The stated mission was to lay out the differences between the two projects to help the public understand the situation. Varmus and Krebs, who had organized the joint press conference, hoped that it would show that the public and private sequencing endeavors were going to work together, or at least weren't going to fight.

But the atmosphere was tense and strained. The NIH leaders worried that a corporate monster might mow down the entire genome program. The timing couldn't have been worse: the congressional appropriations committee responsible for the NIH budget was working its way through the appropriations bill. It was a critical time for funding decisions. Press reports were parroting Perkin-Elmer's claim that Venter's new effort would decipher the genome at one-tenth the cost of the public effort. If congressional leaders believed the Human Genome Project—and by extension other NIH endeavors—could be done faster and cheaper in the private sector, it might seriously threaten the NIH budget.

Standing outside the NIH buildings where the conference was to be held, Varmus nervously awaited its start, wondering what kind of posture White and Venter were going to take. In addition to his budget concerns, Varmus, like Collins, was worried that the information embedded in the human genome sequence might be locked away if the new company took hold of it. Disease research might then be tightly controlled by the private sector, he feared. That wasn't the way he wanted the genome project to work out.

Perkin-Elmer's chief, Tony White, strode toward him. Varmus had never met White before, and wasn't sure how to greet him, especially under the watchful eyes of cameras and reporters. When White introduced himself, White recalled Varmus averting his eyes and being reluctant to shake his hand.[2] White was stunned and embarrassed to be greeted so hesitantly in front of the television cameras. It suddenly hit him that a confrontation, instead of a celebration, might be in the offing.

Inside, however, civility reigned for a while. Both Collins and Varmus proclaimed support for the new endeavor, with Varmus welcoming it as a way to speed up efforts to "get to the goal line" and Collins praising the "significant new initiative." Collins made it clear that his team was not about to give up human sequencing, saying such a move was "vastly premature" given all the uncertainties of the risky venture.

Collins was not even apprised of all the uncertainties. Perkin-Elmer officials had little idea how they were going to finance the new venture—whether they'd keep it inside Perkin-Elmer and finance it out of its earnings, create a totally separate firm, or do something else entirely. The new sequencers didn't yet work. No one had even invented the mathematical tools the venture would need to assemble the human sequence using its proposed method. And Venter's team had not yet developed the procedures they'd need for preparing human DNA for sequencing.

But Collins knew enough to cast doubt on the Perkin-Elmer initiative the day of the news conference. Presumably still fuming from Venter's mouse comment, Collins questioned the whole-genome shotgun tactic, noting it had been "roundly rejected" by government-funded scientists in 1996. He expressed skepticism over whether the data Venter's team collected would be made freely available to researchers everywhere. He underscored the government's policy of requiring its researchers to post their sequence data on the Internet within twenty-four hours of verifying it, implying that this ought to be a universal standard.

Collins was right to defend his program and the importance of continuing it. Taxpayers might rightly be concerned if a project so important were completely handed over to a not-yet-formed firm that was using untested methods and that was first and foremost beholden to its shareholders and not to the general public. Who was to say they would even complete the job?

Many scientists and scientific administrators felt that the public interest would best be served by a joining of private and public forces. The new venture promised a large infusion of talent, money, and ingenuity. The risky methods they proposed were also an opportunity for the public program to reap the benefits of a potentially faster approach to sequencing at no cost to the public.

But if Collins wanted to collaborate with the new firm—and he insisted he did—his remarks on that May Monday did not further his cause. White felt Collins had attacked the integrity of their scientific approach, and he viewed Collins's implicit demand that his company agree to daily data release as unfair, since the firm would not use government funds. The company obviously needed time to mine its data for patentable information first. White felt that his and Venter's compromise solution—to release the company's sequence quarterly—was more than reasonable for a private firm.

When he left that day, White thought, These people are going to do everything they can to hurt us.

Collins and the genome barons weren't the only ones stung by the May 10 announcement. Other genomics companies, such as Genset S.A. in France and Millennium Pharmaceuticals of Cambridge, Massachusetts, were nervous about the potential competition. Perkin-Elmer executives were able to convince company officials that the new firm would be a resource of genetic information for them rather than a competitor.

But Incyte, whose business revolved around selling data, too, was not convinced of that. A day after the announcement, Incyte's chief executive officer, Roy Whitfield, stood up at the BT Alex. Brown Health Care Conference in Baltimore and mocked the new venture. "I wouldn't do that to my shareholders," he said.[3] That same month, Incyte's president, Randal Scott, quipped to *Business Week*, "What they are describing is not a commercial venture. It's really Craig Venter going after the Nobel Prize for sequencing the genome."[4]

The reaction in England was rapid. A week after the announcement, the British medical charity, the Wellcome Trust, announced it was going to approximately double the funding for Britain's Sanger Centre, promising John Sulston a full £130 million ($195 million) for the following seven years. This would finally enable Sulston's team to complete one-third of the genome as Sulston had wanted to do back in 1994. In an article about the Wellcome Trust's plans in *The New York Times*, Michael Morgan, director of genetic research there, was quoted as saying: "To leave this to a private company, which has to make money, seems to me completely and utterly stupid."

In the United States, funding increases were also in the works for the third and final year of the genome project's initial sequencing projects. Of the ten teams then funded for those projects, the groups at Washington University in St. Louis, the University of Washington in Seattle, and Baylor College of Medicine in Houston were awarded the biggest increases. The Institute for Genomic Research had dropped out of the running when Venter announced the competing private initiative—and stepped down as head of TIGR, leaving his wife in charge. Funding for the next five-year phase of the project, in which sequencing would be scaled up, would not be announced until the following year.

That same week in mid-May, leading scientists in the government's sequencing effort gathered at Cold Spring Harbor Laboratory on Long Island, New York, for the 11th Annual Meeting on Genome Mapping, Sequencing & Biology. TIGR's Mark Adams had already been planning to attend the meeting because of his chromosome 16 work. But given the weight and attention of the new announcement, the conference organizers ushered a last-minute invitation to Venter and Hunkapiller. They scheduled the two to speak to a small gathering of about thirty lead researchers from the NIH's main genome centers on Wednesday, May 13, at the laboratory, just before the beginning of the full symposium that evening.

The researchers were eager to hear more about Venter and Hunkapiller's new plan for the human genome. After describing the whole-genome shotgun approach and how his team planned to carry it out, Venter expressed a desire to test his method on a simpler animal first. He didn't want to specify his preferred species, though, because he wanted to talk to his potential collaborators first. After he turned the podium over to Hunkapiller, who was going to describe the capillary machines, Venter walked around the conference table to Gerald Rubin, the eminent biologist coordinating the sequencing of the genome of the fruit fly, *Drosophila melanogaster*, at the University of California, Berkeley.

Venter tapped Rubin on the shoulder and asked to talk to him outside the room. Out in the hall, Venter said that his team would like to sequence the *Drosophila* genome. "Will you work with us?" he asked Rubin. Venter offered to do it for free as a way of demonstrating that their whole-genome shotgun strategy would work, and to make all the data available without restrictions. To Rubin, the offer was too good to refuse. His laboratory had finished only 20 percent of the sequencing so far and would love to receive help with the rest. From the start of the fly genome project in 1992, Rubin had wanted to have the animal's DNA sequence for the sake of biology. But he felt no need to do the job himself. He was not a sequencing jock. He accepted Venter's offer on the spot.

Rubin's eagerness to collaborate with Venter was unusual among genome project biologists. In fact, rumors circulated among researchers in the human genome community that anyone who worked with Venter would no longer be eligible for NIH grants. The NIH funding that Perkin-Elmer had been receiving dried up almost as soon as they announced the new venture. Stanford's Cox, who was mulling over an invitation to join the board of Venter's nascent enterprise, was shocked when Collins told him at the Cold Spring Harbor meeting that Cox could not become involved with Venter's scheme if he wanted to continue working in the public sector. Because of the potential conflict over access to the data, Cox had to choose one or the other, Collins said.

Some members of the genome community also thought Rubin was a traitor to be working with Venter, the man who was trying to steal the human genome from the government and give it to a private company. But Rubin had far less on the line than Cox did. Even if politics stopped the flow of funds to Rubin's lab for genome work, Rubin figured he'd be able to do fruit fly biology, since the fruit fly biologists who would review his grant applications had no axe to grind with

Venter. Rubin was also protected from most of the public sector venom since he was not associated with *human* sequencing. As it turned out, Collins and Varmus encouraged his arrangement with Venter, probably in part because they then could point to the fly collaboration in the halls of Congress if they were called on the carpet for failing to work with Venter on the human genome.

At the Cold Spring Harbor gathering, a new determination to stay the course set in among the publicly funded sequencers. "It is critical that we not retreat from our goal," Collins said. Using the corporate threat as a rallying point, the researchers took turns trashing the quality of Venter's eventual product and warning of the danger that the company would stake a proprietary claim on the genome. Though most of the participants were still in a state of shock from the May announcement, a few, Adams recalled, were outright hostile to him and his boss.

The squabbling continued in *The New York Times* on the last day of the meeting. Washington University's Robert Waterston likened the final sequence Venter would produce to "an encyclopedia ripped to shreds and scattered on the floor." Later in the story, Venter boasted that his team was going to sequence the fruit fly genome "in one-tenth the time" that Waterston and Sulston would take to complete the sequence of the worm *C. elegans*, and that the fly results would also be more accurate.

❧

THE FIRST TANGIBLE hints of how the genome project would respond to Venter's challenge came later in the month at a long-scheduled meeting at Airlie Center in Warrenton, Virginia. There, 120 molecular biologists, sequencers, and database pros gathered to evaluate the proposed details of the project's next five-year plan.

The meeting exposed the first evidence of a crack in the fortress of meticulous completeness that Collins and others had erected until then. Many of those present were not actually producing the sequence; they were interested in using it to answer their own biological questions. And they were suddenly in a hurry for the sequence. Waterston recalls the striking change in the attitude on the part of the "sequence users" from just the year before: "It was clear in that May meeting, that these people didn't care what [kind of sequence] they got, they wanted it *now*." Many biologists were inspired by the scientific firepower that the newly completed yeast genome had given researchers in that field.

Barbara Wold, a top geneticist at the California Institute of Technology, made a strong case that the genome project should respond to the Perkin-Elmer initiative in kind. As chair of the most controversial session of the meeting—how human sequencing should proceed in light of Venter's challenge—Wold recommended that the public project focus on creating a somewhat messy "working draft" of the genome as a near-term goal. As a biologist studying muscle development, she felt that a rough copy of the genome would greatly speed her research, and she'd rather have this version as soon as possible than have to wait for a complete genome. Now, suddenly, with Perkin-Elmer's machines, genome researchers ought to be able to create such a draft in a reasonable time.

Though Waterston and Sulston had proposed something like a working draft of the human genome in 1994, Wold brought the idea back into focus. Still, many of the scientists at the meeting resisted Wold's proposal. Some felt the government-funded project shouldn't be seen as responding in any way to Venter's announcement. Others simply refused to abandon the idea of producing a complete and highly accurate sequence the first time around.

That perfection was still paramount in genome project scientists' minds was clear in a June 9, 1998, *USA Today* article in which Collins said the publicly funded project's Book of Life would contain "every word, page for page, of the full set of instructions for life." By contrast, Venter's book would be far less comprehensive, Collins said: "I think even the *Reader's Digest* is optimistic. More likely this will be the Cliffs Notes or the *Mad Magazine* version."

The differences between the two approaches were underscored at a June 17 hearing on Capitol Hill. Representative Ken Calvert of California, chairman of the House Subcommittee on Energy and the Environment, called the hearing to assess the impact of the Perkin-Elmer effort on the Human Genome Project.

Venter used the hearing to tout the advantages in cost and efficiency his venture had over the public program. People had thought that sequencing the human genome required "billions of dollars" and "decades of work from thousands of scientists," he said. "Now, however, due to new technologies and strategies, the human genome will be accurately and completely covered in one facility by a new company in Rockville, Maryland, with a few hundred workers." Venter preempted critics who forecast his failure by pointing out that he'd proven wrong the critics who'd disparaged his efforts to sequence ESTs and to shotgun the genome of *Haemophilus influenzae*. The critics, he implied, would be wrong again.

Collins was again on the defensive. The $3 billion price tag of the federal program paid for much more than human sequencing, he said. It paid for careful maps of the human genome, including those showing disease genes, for technology development, and for sequencing of the genomes of other organisms. "Up until now, in fact, only a minor fraction of the budget of the Human Genome Project has been devoted to sequencing," he admitted. The federal price tag for complete human sequencing was unknown, but it was unlikely to total anything like $3 billion. (Collins would later estimate the cost of producing an early draft of the genome at about $300 million.)

Collins emphasized completeness and accuracy as the strong points of the federal initiative, but he left attacking the new initiative to Maynard Olson, who was then at the University of Washington in Seattle. Olson predicted that Venter's "approach, as the downside of its efficiency, [would] encounter reasonably catastrophic problems" when it came to assembling the genome from tens of millions of small pieces. He predicted that a hundred thousand serious gaps would remain at the end, seriously compromising the quality of the final product. Olson saw no new science here, or even any significant new technology. "What we have is a press release," he said, referring to the Perkin-Elmer announcement.

Sitting in the audience, Tony White was infuriated by Olson's words. If we throw all our money away with an unworkable approach, what do they care? White thought.

Oddly, even as they crossed verbal swords, Collins and Venter professed a desire to collaborate. Noticing that he and Venter were dressed in similar dark blue suits and blue shirts, Collins quipped that their outfits signaled that he and Venter were on the same team. Sitting next to his rival, Collins said flatly: "This is not a race. We will work together."

Nobody was fooled. *The New York Times* published another story by Nicholas Wade five days later headlined, "It's a Three-Legged Race to Decipher the Human Genome." The competitors were Collins, Venter, and William Haseltine, the head of Human Genome Sciences.

The divergent agendas of the public and private teams made a competition between them inevitable. Collins declared that his team could not work with anyone who did not agree to release DNA sequence data daily as his researchers were doing. His goal was to prevent patents on genetic sequences. But Venter and White thought they were already being generous in agreeing to make their sequences public after three months. Daily data release was out of the question. "Our shareholders would sue us!" White argued. The firm had to have time to mine its data for patentable genes. And to get his patents, Venter had to race the government to find key sequences before its scientists did and dumped them into the public domain.

Ultimately, the public effort may have had more to lose than Venter's company by failing to find common ground. Because they were taxpayers, Venter's employees could access all the data generated by the public project and add it to their own. White put it this way: "We get to see their data; they don't get to see ours. Isn't that sad?" In any race for the genome, the government researchers would have a handicap.

Nevertheless, a dialogue began to develop between the two sides in the summer of 1998. It came in an unlikely form: a meeting between Varmus and White, whom the NIH director had so warily encountered at the early May press conference.

Ever since he'd first heard about the new firm, Varmus had been skeptical of Venter's promise to make his data public, enabling him to deliver the Holy Grail of genetic data to the world. Varmus wondered how such a promise could be made in the context of making a return on Perkin-Elmer's substantial investment. Sometimes, it seemed to Varmus that Venter was trying to please everyone. That didn't seem realistic.

Biologist Arnold Levine, who was soon to be appointed president of Rockefeller University in New York City, suggested that Varmus hear about the business plan directly from White. It seemed like a good idea, and both White and Venter met Varmus at the NIH one summer day.

The new company did not need to sell the raw data, White explained, because there was so much else to sell. His company planned to massage the sequence they churned out into a more functional form—pinpointing the genes, for example, and everything known about their putative functions—using state-of-the-art computer software. They'd end up with various user-friendly data packages to sell to customers ranging from high school teachers to drug industry researchers.

For the first time, Varmus felt he understood the business plan. There seemed to be a way that data could be made publicly available in the context of a business. And with that realization, some of Varmus's distrust melted away.

Varmus also realized that White would be keeping an eye on the more confrontational Venter. White did not condone Venter's tendency to spark catfights rooted in old scientific rivalries. Varmus hoped White would be able to put a stop to that.

18

THE RESPONSE

Collins and his team scrambled to adapt their plans. Adding to the new urgency was the changed mood in the scientific community so evident at Airlie Center in May. After years of opposition to large-scale sequencing projects, academic biologists had finally come to realize how useful a genome at their fingertips could be. For the human genome, quickly pinpointing the genetic script for key biological processes could spawn an unprecedented bounty of information about how biological circuits in the brain, heart, muscle, or lungs were sometimes disrupted, causing disease. That was the first step to finding new cures.

Biologists clearly didn't want to wait for a perfect genome anymore. So Collins was seriously considering the idea of creating a rough draft of the human genome that his researchers could distribute relatively quickly.

But before making such a move, Collins wanted to be sure that the intermediate product the genome project would be creating would be useful. Over the summer of 1998, he asked scientists around the country who used human genetic data—everybody from postdoctoral fellows to computational biologists—to compare the usefulness of more, and less, complete versions of genome sequence. Would they be happy with a rough draft in which long stretches of completed sequence covered about 90 percent of the genome even if these stretches were out of

order and separated by gaps? A more complete version showed a segment still in a few pieces but with the order intact and many more of
the gaps filled in.

In September, the results were in. Without exception, the scientists
said that a rough draft would be very useful. They wanted it.

The rough draft meant significantly less labor for the sequencing
centers for now. To get a polished genome with everything in perfect
order and virtually no gaps required sequencing a human's 3 billion
base pairs at least ten times over, they estimated. That would take at
least several years even with the new machines. But Collins and his
team settled on a draft requiring that every base be sequenced about
four to six times. This was known as four- to sixfold coverage or 4X to
6X coverage. Venter's group had vowed to do a 10X version of the
genome using his riskier method.

In mid-September, Collins and Aristides Patrinos, the DOE genome
project leader, presented the new strategy to the genome institute's
advisory council, a revolving group of fourteen senior genome scientists. The federal labs would now work to produce a rough draft of the
genome by the end of 2001, Venter's target date for his 10X version of
the genome. The publicly funded labs would go on to produce a final
sequence by 2003, two years ahead of the old schedule. Best of all, the
plan was not dependent on a huge funding increase, but on a typical
10 percent annual increase in the genome institute's budget.

One goal of this plan, which the council approved with minor revisions, was to generate data quickly enough to block the patenting of
DNA by companies like Venter's. Nevertheless, in an incongruent
move, Collins had called Venter to suggest collaborating on the new
plan of action.[1] Not surprisingly, the gesture led nowhere.

When the official five-year plan for 1998–2003 was published in
Science on October 23, 1998, it included plans for the rough draft. At
that point, about 6 percent of the human genome had been sequenced.
Costs hovered around a hefty 50 cents per base, a price the plan's
administrators hoped to halve in the next five years. Two of the five
nonhuman organisms whose genomes were to be sequenced under the
Human Genome Project had been completed: the bacterium *Escherichia
coli* and baker's yeast. Still left to complete were the fruit fly, the mouse,

and the worm *Caenorhabditis elegans*. The latter was almost done, and it would be revealed to much fanfare in December.

❧

THE FIVE-YEAR plan did not address the thorniest issue facing Collins. The genome project was then spread out among more than a dozen laboratories, which worked more or less independently. But responding to Venter required a more tight-knit, efficient group of centers. Pushed to the wall, Collins became forceful and decisive. He would whittle down his sequencing network. Anyone could apply to be among the chosen few, but to be a large-scale sequencing laboratory, Collins decided, an institution had to have spelled out at least fifteen million chemical bases of human DNA.

Applications were due in December 1998. A team of peer reviewers evaluated the proposals on paper and, in early January 1999, visited the contenders, considering criteria such as sequencing speed, management, cost-effectiveness, and technological prowess. Each place was ranked against the others. The reports, along with rebuttals of criticisms from the centers themselves, made their way to Collins and the senior scientists on his institute's advisory council.

By February, the winners had been chosen. Robert Waterston's team at Washington University in St. Louis, by far the most productive U.S. sequencing center, with 110 million bases under their belt, was a clear choice. It, along with the Sanger Centre in England, which was Britain's obvious choice to spearhead its genome effort, had the most experience in running large-scale sequencing projects.

Eric Lander's team at MIT's Whitehead Institute, whose forte was robotic factory-like techniques similar to those used for the genome-wide STS signpost map, joined them. During the pilot phase, the MIT squad had also built a small sequencing operation that Lander had brashly claimed—to the disbelief of some of his colleagues—he could scale up by twentyfold or more. The third NIH winner was Richard Gibbs's team at Baylor, which had a good record for innovation and for producing high-quality sequence.

After he'd consolidated his teams, Collins's next task was to get their members to cooperate. By banding together, they could swap innovative sequencing tricks and management tips and share the work in an organized, efficient way. Collins dubbed this new U.S.-British gene team, which also included the Department of Energy sequencers, the G5. But accomplishing such unity would not be easy. The various centers had been rivals for years, jousting for a place in the final assault on the genome, and competitive pressure had long helped push the lab leaders to the top of their profession.

So Collins scheduled a one-day meeting of the G5 in Houston at Baylor College of Medicine in mid-February. Collins wanted to start it off with a bang. It was time to reassess just how quickly his team could get a draft of the genome done.

Two days before the Houston meeting, he sat down to plan. That night and the next, Collins added up the cost and productivity figures that each of the five main sequencing laboratories had provided him. He projected their ability to scale up based on each center's ingenuity and the new automated technology they planned to incorporate into their pipelines. Covering the three billion base-pair genomes five times over, the approximate goal for the working draft, would require thirty million "reads," he calculated, since each read is five hundred bases.

The centers could do that in twelve months. The results were so surprising that he did the calculations again. Twelve months. Collins was stunned. The genome project could complete a draft of the genome eighteen months sooner than he had projected four months ago. This was even faster than Venter said he could get the job done.

Collins knew his plan would require not only a lot of work, but also a difficult change in focus for the centers. Their efforts would have to be heavily weighted toward production, or churning out chemical letters as fast as possible, rather than toward "finishing," or editing and perfecting the product. Collins listed the pros and cons of that approach. To him, the pros were overwhelming. Putting speed over accuracy would bring the greatest benefit to science in the short term. It would also be the best strategy for preventing Venter's nascent firm from locking up vast amounts of basic genetic data and stealing the glory of the genome project.

Suddenly, it was time to leave for Houston. Collins hadn't even had a chance to discuss his latest plan with his own staff.

In Houston, Collins faced his grantees in a small conference room on the fifteenth floor of a gleaming modern tower on the Baylor campus. Perched on turquoise chairs around a gray plastic table were Lander and the host, Baylor's Gibbs, an understated Australian with a boyish face who was proud to show off his team's fancy modern sequencing facility. Also present were the Sanger Centre's Sulston and Elbert Branscomb, director of the DOE's sequencing laboratories. Waterston was sick—he had recently been diagnosed with bowel cancer—and couldn't travel. But he was nevertheless determined to continue working and attended the meeting by telephone.[2]

A faint aura of unease permeated the room, as these ingenious and independent-minded biologists got the sense that a union was being brokered between them. It was awkward to unite with former competitors and hard to put aside pride for the greater good.

The outwardly low-key Washington University team was proud of the sequencing powerhouse they'd built. And while they'd long ago bonded with Sulston's group at the Sanger Centre, with whom they'd been collaborating for years on the worm *C. elegans*, now they had to share their hard work—their secrets—with new teammates.

Collins was outlining his new plan. He drew his matrix of pros and cons on the room's white writing boards. He argued that they restructure their effort to complete the working draft by spring of 2000. He went over the calculations he'd made over the past two sleepless nights. The normally conciliatory Collins got pushy, saying: "We can do this, so why don't we?"

It was truly ambitious. Even the most productive centers, such as Washington University, would have to double or triple their sequencing capacity. The DOE and MIT teams would have to boost their sequencing capacity by more than tenfold in a year to get the job done.

Waterston was seriously worried about whether it was possible. He knew firsthand what it was like to expand a sequencing operation that quickly—the constant hiring, the ordering of new supplies, the reorganizing of the production line. He also didn't want the groups to focus solely on rapidly producing DNA sequence and completely abandon the more time-consuming editing, or finishing, of the genome, as some of his colleagues were advocating. Next spring, he argued, they'd have to turn around and start that editing process in earnest and so could not afford to dump their specialized genome editors and editing technology. He also felt that if they did some editing of the sequence along the way, the process would be easier to manage. When a piece of the sequence was finished, it could be put away.

Other researchers also bemoaned the temporary deemphasis on perfecting their product, since they'd all spent the past year working hard to build up their capacity for doing just that. But everyone agreed that it was important to release as much data as possible as soon as possible to biologists, who wanted it badly. They knew Venter's firm could use the data, too, but at least it could not patent it. And the G5 did not want to be seen as slow, especially in light of the challenge from the private sector. In addition to Venter's company, Incyte had just announced it would be doing significant human sequencing.[3]

At the end of the day, they agreed to the new deadline. To meet it, they would devote the bulk of their new resources to rough-draft sequencing, while continuing to finish the genome at a lower rate. They were united around their common goal—releasing human sequence to the biological community—and their common enemy.

On Monday, March 15, Collins unveiled the plan to the world. "This will have a direct and dramatic acceleration" in the search for new drugs, Collins told *The Wall Street Journal*.[4] "We will be able to use this data to find genes that confer susceptibility to major diseases such as diabetes, common cancers, and coronary artery disease much sooner than previously expected."

The NIH planned to fuel the effort with $86.5 million given to its three largest centers in fiscal year 1999. Washington University received $34.7 million, MIT $34.9 million, and Baylor $16.9 million. Britain's Wellcome Trust also announced an increase in funding for the Sanger Centre for the year, from £16 million ($24 million) to £35 million ($53 million).[5] Meanwhile, the Department of Energy would be increasing its funds for human sequencing from $33 million in FY 1998 to $53 million in FY 1999.

The G5 had shifted into high gear. The genome project would give Venter's firm a run for its money.

19

DEMONIZING A DATABASE

As Craig Venter and Perkin-Elmer executives hatched their secret plans in early 1998, Iceland's Kári Stefánsson was enjoying a similar quiet before being hit by a stormy genetics controversy of his own. As Stefánsson's database bill was secretly circulating through government offices in downtown Reykjavík, his company, Decode Genetics, was celebrating a big breakthrough in the long-frustrating quest to nab a pharmaceutical partner.

Stefánsson had gone on a world tour to attract a business partner but had met with scant success. Many big drug companies had invested heavily in disease gene hunting in hopes of finding new drug targets; they'd turned up precious few good ones, and so they had become somewhat disenchanted with that approach. They favored HGS's approach, which quickly led them to large numbers of genes loosely linked to disease, providing more potential targets.

Some top officials at the Switzerland-based pharmaceutical giant F. Hoffmann-La Roche felt differently, however. They saw disease gene hunting as a direct route toward better diagnostics that could detect the earliest signs of disease. Diagnostics were a priority at Roche. Not every gene discovery needed to produce a good drug target.

On July 3, 1997, *The Wall Street Journal Europe* ran a front-page story on Stefánsson and Decode.[1] It caught the eye of top-ranking

Roche geneticist Klaus Lindpaintner. Lindpaintner had been a genetic epidemiologist at Harvard. He knew that the choice of population was critical to unraveling the roots of complex disease. But the article said Decode was already in talks with potential drug company partners, so Lindpaintner became paranoid that Roche would lose out on the great opportunity in Iceland. Without seeking his boss's approval, Lindpaintner arranged to meet Stefánsson. The next day, in Iceland, he arrived at Decode and saw the pedigrees. He was floored. It was immediately clear to him that the Icelandic population would be hard to beat as a source of genes.

Lindpaintner urged Roche's newly appointed head of worldwide research, Jonathan Knowles, to take a hard look at Decode and meet with the charismatic Stefánsson. Knowles was also impressed—by Decode's gene-hunting strategy and its technology. He was quite surprised to find a state-of-the-art, extraordinarily cost-effective biotech company in Reykjavík.

After several months of negotiation, Roche and Decode unveiled a deal valued at up to $200 million in February 1998. The five-year agreement was one of the biggest drug company–biotech deals in gene research, topping the $125 million HGS–SmithKline Beecham deal. Roche received a small equity stake in Decode (less than 10 percent) and exclusive worldwide rights to all drugs and diagnostic products resulting from Decode's research.

Decode was to direct the new funds toward finding genes for twelve diseases: four cardiovascular diseases, four nervous system ailments, and four metabolic disorders, including adult-onset diabetes. To win the cooperation of Icelandic citizens in the research, Roche pledged to give away any drugs developed from Decode's findings to Icelanders for free.

The influx of funds fueled a rapid expansion at Decode. The firm bought the building it had been renting and spread out over all three of its floors. Stefánsson filled the bottom-floor laboratory with new equipment, quadrupling the number of DNA-copying robots and tripling its DNA-copying capacity. That would allow Decode researchers to pinpoint a thousand little landmarks in each person's genetic code instead of a few hundred, giving them a higher resolution view of the genome, which they expected would greatly improve their ability to find disease-causing genes. It wasn't long before Stefánsson had carved out gene-fishing jobs for about 250 people, nearly one of every thousand people living in Iceland.

The glow of Decode's new wealth had hardly dimmed when on March 23, the Icelandic Parliament, Althingi, called a meeting with Iceland's Director of Public Health (akin to the U.S. Surgeon General), Ólafur Ólafsson, and a handful of top Icelandic medical officials. As a political formality, they wanted these eminent officials—which included the heads of the Icelandic Medical Association and the Icelandic Psychiatric Association—to give their nod to Stefánsson's idea of a bill to create a health sector database.

Instead of a nod, they got an earful. Ólafsson and the other doctors were incensed. They were particularly disturbed by the fact that the bill did not require that patients be asked for their consent to participate. Instead, every Icelandic citizen was assumed to have consented to the release of his or her own health care data unless the person deliberately "opted out" by signing an official form. An entire nation's health records were going to be handed over to a private firm without asking permission from the individuals involved. The bill was moving quickly. Althingi planned to rule on it within weeks.

Over Ólafsson's protests, the bill was submitted to Althingi on March 31. On the same day, the bill's text was posted on the homepage of the Ministry of Health and Social Security. That's when the storm broke out.

As word of the bill spread, many doctors and scientists read it on the Internet and became alarmed. Among the first were a soft-spoken psychiatrist named Pétur Hauksson and an amiable biochemistry professor at the University of Iceland named Sigmundur Gudbjarnason. Both were outraged by what they regarded as the bill's violations of personal privacy and began mobilizing doctors, scientists, and citizens to oppose the pending legislation. On behalf of the Icelandic Mental Health Alliance, which Hauksson led, the psychiatrist fired off letters to patients' rights organizations around the world asking them to put pressure on the ministry to withdraw the bill. He also sent a formal opinion against the bill to the parliamentary committee handling it.

Because of the sensitivity of health records, Hauksson and Gudbjarnason argued, the law should require that patients give consent before the release of any personal data. Not everyone would be aware of his or her right to opt out of the database, fully cognizant that his or her medical data were being submitted to it, or even physically and mentally able to opt out. Hauksson argued that people who could find the form, understand the form, and find a stamp were safe. But anyone without those abilities was unprotected.

Gudbjarnason worried about the ramifications of giving out medical information. He felt that if, say, the private confessions about a relationship got into the wrong hands, that knowledge might break up a marriage. Other information, such as venereal disease status, might humiliate people if it ever got out. Hauksson felt that patients wouldn't be able to trust doctors if they knew that doctors would send their information to a private company that would use it in unknown ways. He felt this was a breach of the Hippocratic oath.

According to the law, doctors in private practice would have the right to keep their patients' records private. But in hospitals, physicians would have to abide by the decision of the hospital's board about whether to contribute to the database. Some said they would not. Hauksson, who has his own practice but is also affiliated with a hospital, vowed to break the law. He promised to hide his records and, if all else failed, move to the United States. Other physicians were less outwardly defiant but equally inclined to disobey. One planned to submit only less sensitive patient data, thus appearing to abide by the law while keeping more inflammatory information confidential.

Aside from privacy, Icelandic scientists were angry because they felt the bill would give Stefánsson and his team a monopoly on genetics research in Iceland. Decode would have full access to the entire country's health records and exclusive rights to the data they gleaned (separately) from much of the population's DNA. Many felt that this was too big a prize to give a single company, even if it had paid handsomely for the privilege.

Within two weeks of the bill's submission to Althingi, Iceland's most influential medical society, the Icelandic Medical Association, wrote to the Ministry of Health and Social Security to protest the document. On April 24, Ólafsson, Hauksson, and other public health leaders personally complained to ministry and Althingi officials in a second meeting. In response to the outcry, the ministry withdrew the bill for revisions within a month of its public posting.

But the cacophony of voices protesting Stefánsson's proposed database only got louder. One hundred and eighty of the nation's nine hundred doctors signed a petition refusing to cooperate. Some seven hundred articles on the controversy—many, if not most, critical of the bill—appeared in the three newspapers in Iceland that year. Town meetings were held all across the country to discuss what became perhaps the most debated piece of legislation in the history of the republic.

The attacks exhausted, angered, and depressed Stefánsson, not least because the survival of his company rested on the support of the Icelandic people. Without it, Icelanders could shut the company down by simply refusing to participate in its research. Stefánsson worked tirelessly to keep the people under his wing. Hardly a day passed when he didn't give speeches or interviews to journalists, promoting his research and defending his database.

No collection of Icelanders was too small or insignificant to escape Stefánsson's notice. On a dreary Saturday, he ascended the stairs of a gray building in the oldest section of Reykjavík to address a small gathering of people with Tourette's syndrome. These people emitted unusual sounds and engaged in purposeless movements called "ticks" that they could not control.

The audience, perched on metal folding chairs in the small, dim room, greeted him warmly. Stefánsson spoke in Icelandic with his usual intensity, his brow furrowed, his tone emphatic, and apparently unfazed as one rapt listener or another discharged a popping noise, cleared his throat, or repetitively pursed her lips. The audience was transfixed.

He reassured his listeners that they were free to choose whether to participate in genetic studies, that nobody would take their blood without their consent. He occasionally cracked a wry smile and, at one point, the room erupted in laughter, apparently from a joke he told.

To group after group, reporter after reporter, Stefánsson argued his case in a kind of grassroots political campaign. He countered charges that his firm would have exclusive rights to use the database by explaining that the Icelandic health care system would also be able to use it, as could Icelandic scientists to do noncommercial research and foreign scientists who collaborate with Icelandic scientists. As for privacy, Stefánsson argued that researchers performing population-based studies of disease have for centuries relied on implied consent when analyzing medical records for statistical trends. The reason was simple: it was impractical, if not impossible, to chase down everyone listed in the records.

In any case, the database law already included privacy protections. It called for encrypting the identities of everyone—both in the hospitals and clinics and again upon arrival in the central database. The database would also include administrative locks against unauthorized use and software that prohibited any user from obtaining information about a single person, or even a group of fewer than ten people. Stefánsson vowed that information collected in the database would be safeguarded as well as any data in the world, although nobody pretended that the security system could block a determined soul from deducing a particular Icelander's identity using the available health and genealogy clues.

Stefánsson's campaign was effective. He had an unusual charm that quickly endeared him to people. His charisma was something his foes feared, because they knew it could sway—in their eyes, fool—a nation. Stefánsson's magnetism stemmed partly from his willingness to express his feelings, even to near strangers, bringing them closer to him. But he may have also learned how to capture people's attention from his father, a radio journalist whom Stefánsson remembers as a great storyteller.

Stefánsson did not just use his charm on the public. He also used it on politicians in the Ministry of Health and Social Security and Althingi. And with it, he breathed new life into his health care database bill.

On July 31, the ministry unveiled a revised bill with additional privacy protections and circulated it among dozens of organizations for comment. Anyone could ask that his or her data be coded in such a way that it could never be decoded, although it was unclear how such protections would still allow new information about the same person to be added later. The bill also increased the government's oversight of the database project, but the bill's critics felt the oversight committees would be relatively powerless.

As the new bill was put before the Althingi again on October 9, a new organization composed largely of doctors and scientists sprouted up to demand informed consent for any information submitted to the database, among other stipulations. The group called itself Mannvernd for "protection of man," and chose the diplomatic Gudbjarnason as its first leader.

In the final weeks before the vote in Althingi, the campaign against the bill spread overseas.

In the United States, Mary-Claire King, a geneticist at the University of Washington, and Henry Greely, an expert on genetics and the law at Stanford, wrote a letter to Prime Minister David Oddsson and other Icelandic ministers urging them to reexamine the database plan. They were troubled both by what they considered the bill's lack of respect for individual rights and by the prospect of a de facto monopoly of genetics research in Iceland. And Richard Lewontin, a Harvard geneticist, chimed in with a letter to an Icelandic newspaper expressing similar concerns but adding the more inflammatory suggestion that researchers worldwide stage a scientific boycott of Iceland to protest the bill.

Stefánsson was furious. He thought the attacks were prompted by the misinformation spread by his opponents in Iceland—in particular, the rumor that people's DNA would be handled the same way as more traditional doctor's records. In fact, the genetic data would be gathered separately and only from those who volunteered to donate blood and signed a consent form permitting Decode to use it. Over and over again, Stefánsson repeated this mantra only to see the issue misrepresented once more in foreign newspapers or TV shows. One night, he listened to Stanford's Greely on ABC's *Nightline* speak enthusiastically about the scientific promise of the database—but criticize the gathering of information without informed consent, implicitly lumping genetic information in with the health care data.[2]

It was exasperating.

Most serious critics of the database, including Greely, did understand that genetic data would require informed consent, but they still had concerns. For example, when each person donated his blood, that person would be permitting the use of his DNA in any study of a disease that impacted him or his family. Some experts worried that people could not knowledgeably give such blanket consent because nobody could foresee everything the DNA might be used for in the future.

Stefánsson thought this concern was unfounded. People of reasonable intelligence could be fully informed and, in giving their consent, were merely exercising their right to self-determination, he contended.

For the most part, Icelanders wanted to exercise that right. They were excited about the potential benefits of the database and eager to help make it happen in the hope that the research might benefit their descendants if not themselves. They were willing to trade some privacy for progress, and were apparently not terribly worried about handing over their health records to a private firm. If Decode did not keep its promises to the people of Iceland, it would probably be disciplined if not shut down.

By the parliamentary vote on December 17, 75 percent of the population supported Stefánsson's bill. It passed in parliament by only a slightly smaller margin: thirty-seven votes in favor, twenty against, and six abstentions.

But the controversy raged on, as if inflamed by the bill's passage rather than calmed by it. Hardly more than a month had passed before Harvard's Lewontin condemned the bill once again, this time in *The New York Times*. In an opinion piece called "People Are Not Commodities" published on January 23, 1999, Lewontin suggested that Icelanders had been misled about issues such as the need for informed consent in medical research. Because he felt most Icelanders would not have the foresight to opt out of the database, he wrote, "The great bulk of the population will be tools of Decode and its backers."

By early May 1999, after a token competition with one other firm, the Ministry of Health and Social Security picked Decode to run the database. Stefánsson and his team then began negotiating with the government over technical and procedural details before they could receive the official license to get started.

At the same time, Mannvernd launched a campaign to encourage patients to opt out of the database. But popular support in Iceland for the legislation continued to grow. Another poll taken three months after it passed showed that support for it had risen to 88 percent. Stefánsson took heart in this news, since he reasoned it would be extremely difficult, if not impossible, for recalcitrant doctors to imperil the plan against the will of 88 percent of the nation. By June, only seventy-four hundred Icelanders (or slightly less than 3 percent of the population) had followed Mannvernd's advice to opt out, and Mannvernd geared up to send the official opt-out forms to every home in Iceland, though, in the end, it lacked the funding to do so.

Though Stefánsson bristled at the nonstop criticism, in more reflective moments he was in favor of having a serious debate. "It is quintessential that you have critics when it comes to an issue of this sort. The critics keep those who execute an idea on their toes," he said. "It's dangerous for democracy when a complicated issue is brought forward without any opponents."

⁂

WHILE THE DATABASE controversy raged, Decode researchers soldiered on with their genetic studies, studying the blood of groups of patients who agreed to participate. They matched this genetic data with those patients' diagnoses and with genealogical data, which they didn't need a bill to collect. The only data they didn't yet have at their disposal were the health records collected by doctors in clinics and hospitals throughout the nation. Incorporating those records into their work would have to await not only Decode's official license to use the health care database, but also a substantial internal effort to build the electronic infrastructure for it.

Meanwhile, each of more than thirty Decode research teams worked on tracking down a gene for a particular disease. One team searched for a gene for osteoarthritis, the most common form of arthritis in which the bones' cartilage cushions wear away. With the new *genealogy* database—that is, the electronic family trees Stefáns-son's team built with computer programmer Friðrik Skúlason—Decode researchers could quickly select a vast number of patients with the familial relationships needed to study each disease.

In the case of hip osteoarthritis, collaborating orthopedic surgeons from three Icelandic hospitals compiled a long list of nearly three thousand patients with arthritis so severe they'd needed surgery to replace their hip joints. Iceland's Data Protection Commission, a government body devoted to the protection of privacy, then disguised the patients' names with numbers and sent them to Decode. Decode scientists ran those lists through their computerized family tree, in which people's identities are represented by the same codes used to encrypt the patient lists.

In minutes, the family connections of all the patients, with disguised identities, popped up on the computer screen. The researchers thus got an instant family tree without having to ask the patients their family histories or spend months traveling from church to church and piecing together written records.

Decode statisticians and project leaders selected about eight hundred anonymous patients from the tree based on their family relationships. They sent this list to the Data Protection Commission for decoding. The list of resulting names then arrived at the offices of their doctors, who asked the patients to participate in the study. Those who agreed gave blood.

When the coded vials of blood arrived at Decode, scientists working in a small room with a view of the sea and snow-patched mountains extracted DNA from the blood. They then ferried the opaque blobs of genetic material downstairs to a large room that resonated with the whir of machinery, including a row of twenty-two DNA-copying robots. There, scientists began the process of looking for chromosomal landmarks that heralded the presence of an inherited osteoarthritis gene.

By spring 1999, the osteoarthritis team found the general location of a gene for the disease in these Icelanders, one they quickly narrowed to a region containing less than ten genes. It was the first find the company had made under its collaboration with Roche—and the biggest triumph so far in Roche's genetic research program, giving the drug firm potentially powerful genetic firepower to develop drugs and diagnostic tools for the disease.[3] That same spring, the Human Genome Project launched its accelerated initiative to sequence the entire human genome, the spoils of which Stefánsson planned to someday incorporate into his database.

Stefánsson saw benefits of his work aside from helping to develop brave new pharmaceuticals. Genetic knowledge, as he saw it, could help people live better and longer by protecting themselves. He envisioned people getting their genomes scanned for genetic quirks that predisposed them toward, or protected them against, various diseases. Such a scan would not usually forecast fate. Instead, it would tell people where they were most vulnerable so they could make lifestyle choices that could reduce these vulnerabilities. For example, if a person discovered he had a mutation in a gene that leads to emphysema if he smokes, he would have an added incentive not to smoke. Or if someone had a mutation in a gene predisposing her to alcoholism, she would be advised to steer clear of alcohol.

Knowing one has a gene for osteoarthritis might inspire preventive behavior such as taking off excess weight, reducing pressure on the joints. And a predisposition toward other conditions, such as cardiovascular disease, could influence some people to alter their diet or monitor the health of their hearts more diligently. When Decode scientists localized the gene for preeclampsia, they could hope that the finding might someday be used to identify those women who are at risk for the skyrocketing blood pressure and associated havoc the disease wreaks during pregnancy. Those women could take precautions and perhaps preventive medications early in their pregnancies.

But beyond such actions, Stefánsson had more philosophical reasons for pursuing his work. "What we are doing is increasing the knowledge man has of himself. Knowledge for the sake of knowledge is always important. It makes you into a better man, and us into a better species," he said. "I want to believe that it is important for us, no matter whether it turns into advances in medicine or leads to lifestyle changes that will prolong our lives."

❧

STEFÁNSSON WAS ALSO intensely curious about what influenced the human life span. He knew that diet and exercise had a strong effect on how long people live. So did access to quality medical care, sanitation, and a huge host of environmental factors. But Stefánsson wondered about the existence of genes that predisposed some people to live well into old age.

His interest was almost morbid, since his parents' early deaths provided little hope that he had inherited anything that would predispose him to a long life. Nevertheless, he decided he wanted to know whether there was a longevity gene.

So one summer Sunday afternoon, Stefánsson sat at his computer and started fishing for one. For longevity, he realized that he already had a nationwide data set. His firm's genealogy database then contained the birthdays and times of death for six hundred thousand Icelanders, most of those who had ever lived. Using the database, he and a medical student generated a list of fifteen hundred Icelanders who lived to be more than ninety years old and identified their family relationships. They discovered that the people who had lived the longest not only clustered into families, but also were, as a group, more related to one another than were same-size groups of people who had died younger.

The result showed that longevity ran in families. But then Stefánsson wanted to know whether genes, as opposed to shared environmental factors, governed the trait. So he asked his database this question: If a person has a sibling who is over ninety, how likely is that person to become ninety compared with someone without a sibling who is so old? The computer spit out a statistic. Stefánsson then explored how this figure changed as he walked back through a family's generations. The data began to suggest the influence of one or a few major genes.

Stefánsson was surprised to find that such a simple genetic mechanism could govern longevity, given all the things that could abolish its influence. "It doesn't matter if you have the longevity gene if you step in front of a truck," Stefánsson pointed out. "If you have a mutation in the breast cancer gene and you develop breast cancer, you could die in spite of having the longevity gene." For a gene's influence on longevity to be felt in the face of so many other genetic and environmental factors, it must be enormously strong.

The finding hardly augured well for Stefánsson's chances of becoming an old man. But he had little time to ruminate as he became sucked into the boiling controversy over his proposed health care database. This political matter forced him to drop the longevity project for nearly a year. Then, in mid-1999, Stefánsson received approval to collect blood from elderly Icelanders so he could continue to pursue his gloomy obsession by directing a concerted search for the longevity gene.

Stefánsson had little inkling of what kind of gene he was looking for, and even less of an idea of what he would do with the gene once he found it. While he had few doubts about helping people live healthier lives, he was far less certain of the value of extending human life in general. As a doctor, when he sat in front of an individual, he wanted to do everything he could to lengthen that person's life. But he didn't have the faintest idea whether it would be beneficial to extend the life span of the entire human race.

Still, Stefánsson was encouraged that he had been able to detect the presence of the longevity gene using statistics alone, since that suggested that a nationwide health care database could do the same for specific diseases. He imagined using its disease information to examine not only the causes of individual diseases but also how one disease might lead to others. Obesity, for example, is a risk factor for various ailments including heart disease and diabetes. The longevity work clearly demonstrated the importance of having a large body of data.

And in this large body of data, Stefánsson also saw a way to better health care that did not involve new drugs or chemical diagnostic kits. He envisioned a tool for storing and processing patient information that could help doctors arrive at better diagnoses and treatments.

Stefánsson often talked about what he deemed a lack of technical sophistication among medical professionals, a view that probably contributed to the tension between him and Icelandic physicians. In particular, he felt that using the doctor's brain as the main way of storing and analyzing patient data was a primitive way of making life-or-death medical decisions. Any one doctor's knowledge is incomplete, even if he or she could access all of it, and human brains are hardly perfect at analyzing data.

"You don't go to a bank without the help of software," Stefánsson argued. "And yet we still live in a world where medicine is performed as art. That means doing it in a manner people cannot justify or under-stand—which is totally unacceptable." Computers, of course, could store and analyze vastly more data than any person. While people had been trying, without great success, to develop medical "expert sys-tems" for more than a decade, Stefánsson thought he could succeed with this now. Iceland's health care database would provide excellent fodder for designing such software.

The database, as he saw it, would contain more data and more precise data than anyone had compiled in the past—in particular, information about genes. Instead of just trying to match a patient's physical symptoms with those characterizing a disease, a computer could be given a patient's genetic profile. It could compare that profile to the genetic quirks known to characterize various diseases. The best match would be the diagnosis.

In addition to improving modern health care, Stefánsson envisioned his new software improving the care of people in poor nations by deliv-ering medicine's most analytical aspects over the Internet. In other words, the medically disenfranchised might someday receive expert diagnoses and treatment decisions from their computers.

Stefánsson's vision was a natural extension of the use of powerful computers to store and analyze vast quantities of genetic information for biological research. Stefánsson was aware of the public data banks starting to fill with DNA letters decoded by the scientists in the Human Genome Project. Biologists around the world were tapping these emerging data banks for information pivotal to medically useful discoveries. But no one had greater ambitions for the use of comput-ers in gene-based medicine than the scientists who were about to ask them to mold a human genome from seventy million minuscule slivers of DNA.

20

THE MAKING OF A MONSTER

Arizona computer scientist Eugene Myers had lost hope that anybody would ever test his and Marshfield geneticist James Weber's one-step, shotgun-sequencing strategy on the human genome. He didn't even see the May 10, 1998, article in *The New York Times* trumpeting Perkin-Elmer's new project. The buzz caught Myers's attention only a week later when he received an e-mail from a colleague asking if he knew about Perkin-Elmer's plans.

Curious, Myers called his friend Granger Sutton at TIGR, who confirmed the news. It took a while for Myers to realize that Venter's team wasn't just planning to sequence the human genome, but to shotgun it, testing the method he'd been painfully pushing for years. Surprised at his own words, Myers found himself asking Sutton whether Venter might be interested in his joining the team.

Myers loved the desert, his peaceful life, and his many friends in Arizona. He'd given up other great career opportunities so he could stay put. But now, Myers had something to prove. He believed in this and wanted the chance to help make it happen. He knew that he would wonder the rest of his life if he gave up that chance.

Sutton was encouraging, so Myers jumped on a plane to Washington to meet with Sutton, Venter, and various other members of the small group that had assembled to start the new venture. They seemed interested in him. But the next month, while Myers was in Europe working with colleagues on collaborative projects, he began to wonder whether joining the new company was right for him. Was his own ambition tripping him up? His doubts only grew as for several weeks nobody from TIGR called him to follow up.

But the nascent team of Venter's firm—all of them still working at TIGR—had been distracted by the crush of media attention, the attacks from the public sector, and the need to find office and lab space for the new company. They eventually rented a white, blue-trimmed office building in Rockville, Maryland, that had been vacant for about eighteen months. It looked rather sordid, and it smelled musty. While construction crews began renovations, temporary quarters were established down the road at TIGR for the company's first labs.

They were also in the midst of a recruiting drive. Using some of the $338 million in cash Perkin-Elmer gave the new firm at its launch, the company set out to hire software engineers and biologists of various stripes. On June 27, a dozen human resource specialists flew down to TIGR from Perkin-Elmer's Boston area unit to hold a job fair.

Venter's new company had neither a building nor a name, but by nine-thirty A.M., the line of waiting applicants extended out the front door. The job seekers ranged in experience from postdoctoral fellows hunting for permanent posts to top scientists from biotechnology firms. About three hundred people came to the fair, apparently taken with the promise of human sequencing and the hype over the new endeavor.

Myers had not been forgotten. When he returned from Europe, Venter called him with an offer. Still somewhat reluctant, Myers held out for good terms—a sizable salary increase and stock options—and got them.

Within two weeks, Myers packed up his Arizona office, and its eighteen years worth of files and memories. Looking at the empty space, he was amazed he'd wrapped up so much of his life so quickly. He then set out to drive across the country alone, with little idea of what was in store for him.

Myers's new employer soon had a name. It was called Celera Genomics to suggest "celerity," meaning swiftness of action, or speed. Celera is also a string of letters embedded in the word "acceleration."

Myers arrived at Celera's new, rented headquarters—a still largely vacant tower infested with construction crews—just after the first twenty or so employees had moved in during the first week of August. In his makeshift office with its rented furniture, Myers felt a little disoriented. He missed the books that filled his old office and, ironically, didn't know how to use the main appliance in his office—the computer—which ran Windows instead of Unix, the operating system he'd always used. His work was constantly interrupted by construction noise and power outages.

Nevertheless, he was thrilled to be there. There was an "anything-it-takes kind of atmosphere," he recalled. "We were all willing to camp out."

Myers's team—for the moment, mainly himself, Sutton, and a computer scientist named Art Delcher—got together in Myers's second-floor office to figure out how, exactly, they were going to assemble the human genome from lots of little sequenced bits. Myers had shown it could be done with the simulator he'd built in Arizona. The main hurdle was finding a way of detecting repetitive DNA, and he'd even taken a semisuccessful crack at that.

But what they didn't know for sure—what kept Myers up at night—was whether they could do it as rapidly as Venter needed it. They had to assemble the fruit fly genome as a test of their method, followed by the human genome, in less than two years.

Scribbling on a whiteboard and eating Chinese food, the threesome settled upon a strategy. They didn't have time for much wholesale experimentation, so they decided to break the problem into simpler stages that they would validate one at a time. When all the stages were complete, they would work to transform the raw strips of DNA sequence coming off the sequencing machines into an assembled genome.

Celera wouldn't even have any sequencing machines until December, when the first of Perkin-Elmer's PRISM 3700s were scheduled to roll off the delivery truck. So Myers worked on a simulator, an upgraded version of the one he'd built in Arizona. Using an old Toshiba laptop he'd lugged with him from Arizona, he created synthetic genomes with repetitive DNA sequences like those he expected to see in the human and fly genomes. He used these virtual genomes to test the assembly software before real data became available.

As he worked on the simulator that fall, Myers also was flying around the country meeting with various computer vendors—IBM, Sun Microsystems, Silicon Graphics, and others—to help find somebody to supply the big computers his team would need to run its programs on a large scale. For his first six months at Celera, Myers worked all the time, barely stopping to eat and sleep.

⚮

AT CELERA'S TEMPORARY labs at TIGR, a team of biologists was trying to set up the operation that would produce the DNA data Myers's crew would assemble. Through the fall, Hamilton Smith, the Nobel laureate who dreamed up TIGR's first bacterial-sequencing project, worked by himself at a small bench. He was developing a new system for cloning fruit fly DNA so the Celera team could start its test project on the fruit fly. By December, he'd completed and checked his work, just in time for the first fruit fly DNA to arrive from California.

Meanwhile Venter's longtime collaborator, Mark Adams, and his team of researchers were trying to set up the laboratory assembly line that would marshal Smith's library of molecules through the many steps required before the DNA-sequencing machines could take over. The process itself was fairly standard. First, they would separate Smith's clones, each of which harbored a fragment of the fly's genome, by pouring the whole lot onto square plastic plates coated with a straw-colored substance on which the bacteria could grow. The bacteria would then form separate colonies, which appear as thousands of white spots on the plates. Then a robot would scoop up the spots and deposit them into plastic plates with ninety-six tiny pits. After being copied still further, the pieces would undergo the sequencing reactions invented by Frederick Sanger.

Adams's big challenge was to make the procedures work on an unprecedented scale. If Celera's army of machines was going to sequence two hundred thousand samples a day, the company would need a DNA-preparation process that could feed those machines two hundred thousand DNA fragments a day. This was forty times the number TIGR was processing at the time.

Back when they were secretly planning the project, an Applied Biosystems scientist had promised Adams that his group had an entire suite of new miniaturized, automated technologies for preparing massive amounts of DNA for sequencing. But when Adams looked into implementing these tricks, he discovered that not a single one of them was ready. It was suddenly up to him to figure out how to scale up the DNA-preparation process using more traditional methods. That summer and fall, he gradually found solutions, adapting the methods they'd used at TIGR for the new project.

As the new year dawned, Venter and Berkeley's Gerald Rubin formalized their agreement to sequence the fruit fly genome. Celera researchers would churn out the sequence of three million tiny fragments of that genome. Since Celera's whole-genome shotgun strategy for sequencing the fly was untested, the Berkeley group and a team at Baylor College of Medicine planned to construct a map of the genome as a backup. If the untested strategy failed, the map could help Myers's team assemble the sequence fragments into a complete genome.

The teams would also jointly polish the sequence and interpret it, a process known as annotation. Celera promised to release its preliminary fly data by July 1999. Many scientists then hoped this arrangement would serve as a model for a similar partnership on the human genome. With Celera's help, the fly genome would probably be completed two years ahead of the original schedule and save the government $10 million. It would unearth a treasure trove of genetic data on a creature that had been the subject of intense scientific interest for nearly a century. The genome sequence promised to help researchers interpret humanity's genetic instructions when they deciphered them, since fruit flies and humans share the same evolutionary history.

But the Celera team would not start sequencing for several months. In January, Smith and a new assistant named Cindy Pfannkoch started making the fly DNA libraries, the first step in preparing the DNA for sequencing. They cracked open the small conical plastic tube that held the raw or "naked" fruit fly DNA shipped from the Berkeley lab in a squirt of liquid. They burst the DNA into fine drops using a device like a perfume sprayer, breaking the DNA into pieces, some of them ten thousand base pairs long and others two thousand base pairs long. (The plan was to sequence both ends of each piece, with the sequence from the larger pieces providing information for ordering and orienting the smaller ones.) Then they cloned the pieces in bacteria.

By the end of February, the libraries were done, and Adams's team—now working in the basement of the rented Celera building—was putting the finishing touches on the procedures for running samples of DNA through the laboratory. But the sequencing factory, housed in the basement with Adams's team, wasn't yet ready. All of the sequencers hadn't arrived, and the ones that had kept breaking down, while the technicians who were supposed to fix the machines were still being trained.

It wasn't until the end of April that Celera's first permanent sequencing center, on the fourth floor of its new building, was ready. Celera had a hundred sequencers by then. Not all of them worked, but it was time to get started.

❧

ON MAY 5, 1999, Venter's company started trading on the New York Stock Exchange under the symbol CRA. Its parent company, Perkin-Elmer, whose name Tony White had changed to PE Corp. (and later to Applera Corp.), had sold its analytical instruments business and launched Celera Genomics as a separately traded "tracking" stock on the NYSE. Celera and Michael Hunkapiller's Applied Biosystems unit remained under the same corporate roof with a single board of directors, CEO, and executive committee. But the two stocks traded independently.

White wanted the companies to remain as closely tied as possible so they could collaborate with little hassle. Someday, White envisioned the two companies selling products—such as personal gene analysis—that combined Applied Biosystem's machines with Celera's gene databases. But at the same time, the two firms could have separate groups of shareholders, with more risk-averse investors choosing the more established firm, Applied Biosystems. It seemed like the best of both worlds.

Celera's stock opened at $24 per share. Three weeks later, it closed at just $16 per share.[1] The initial drop of the stock price was undoubtedly due to the selling of Celera's shares by PE's conservative shareholders, who inherited shares of both tracking stocks. But Celera's business plan also seemed murky to many investors, who were uncertain about how the company was going to make money from the raw DNA data it planned to generate.

All this probably made White irritable, and he became irate when Venter began fueling the feud between his firm and the Human Genome Project in the media. On May 18, Venter was profiled in *The New York Times* under the headline, "The Genome's Combative Entrepreneur." In the article, Venter disparaged Collins and Britain's Michael Morgan, claiming they were "putting good money after bad." "Their plan is not to finish the genome," but to prevent him from patenting it, he said.

White didn't just feel that Venter's bad-boy, rock-star persona was inappropriate for a president of a publicly traded company. White was a bit of a character himself, a man who makes it known that his cousin was the drummer for the popular rock band Bon Jovi. But Venter's acrimonious declarations in print were bad for business.

Was Venter really interested in becoming a businessman or was he, as some of his competitors claimed, simply going after the Nobel Prize? The more Venter squabbled about the science, the more credible the claims seemed that Celera was about scientific accolades rather than making money. After the story appeared, White started getting messages on his answering machine indicating that large holders of Celera's stock were getting cold feet about the company. He decided he had to tame his new underling.

One day Venter gave him the opportunity. He innocently asked White what he thought was wrong with the stock.[2] "Craig, I think it's you," White exploded. "I think you need to . . . shut up!"

But suppressing Venter wouldn't be easy. Venter had long made a habit of trumpeting his victories on the front pages of newspapers and the covers of magazines. He lived the media high life, and he plastered the walls of the waiting area outside his Celera office with framed press clippings about himself, including front-page profiles in *The Washington Post* and *USA Today*. The articles spilled out into the hallway and even decorated the main lobby of the building, where a two-page spread titled "Gene Maverick" from the January 11 *Time* had been on display for months. Visitors were greeted by an image of Venter in a white lab coat, standing before a long hallway glowing eerily under a yellow-green light. He bore a look of exaggerated seriousness like a cartoon character unveiling a plot to take over the universe.

For his numerous media photo shoots, Venter had several fetching images, including a mock quizzical "who me?" look, a sci-fi stare, and a cunning Cheshire cat grin. White didn't so much object to Venter attracting media attention, but he wanted to make sure Venter came across as a proper businessman, a transformation so difficult that White likened it to adapting to the diagnosis of a terminal illness. First came denial—boss or no boss, Venter would do what he wanted. Denial would be followed by anger, then acceptance.

In July 1999, Venter seemed to be in the anger stage. "While other people may think there's a lot of fame and glory in what we're doing, on a daily basis it's mostly harassment and petty shit that you get from people," he complained. "I now have a thousand bosses at least, or ten thousand bosses," he said, referring to his stockholders with a tinge of both pride and dismay. "Even some of the smartest people in some of these organizations have very short memories. It doesn't matter that all this has been built in the last eleven months, and nobody's ever done anything like this before. It's still: What have you done for me lately?"

By fall, however, Venter seemed to have begun to accept his new toned-down role. White was pleased. "If you look at Craig's public statements today versus what they were three or four months ago, it's two different people," White said. "He's matured in this area, and investor confidence goes along with that." Venter's new more professional demeanor apparently worked wonders for the share price. On September 22, for example, Celera was trading at $47 per share, up from $17 in mid-June, raising the total value of PE Corp. by $900 million since May.

❦

WHILE CELERA'S MAIN building still had the feel of a construction site, eighty sequencers already hummed noisily in a huge room, processing the carefully prepared strips of fruit fly DNA. The floors had been reinforced to support the massive weight of the machines, and special silver-colored ducts had been installed to vent out of the room the enormous heat from the sequencers. Even so, the room required extra air-conditioning.

Each sequencing machine held four silver-colored trays, each housing 384 samples in separate little wells, keeping the machines busy for twenty-four hours without human intervention. A robot automatically transferred material from the plates to a tiny trough on the machine that fed a fan of a hundred auburn-colored hair-thin glass tubes. Each tube, or capillary, took up many copies of a unique piece of fly DNA. Under the influence of an electric field, the pieces swam through the honeylike goo that filled capillaries, with the smaller pieces swimming faster and popping out first to be recorded by a detector at the exit. From the size of the pieces and the color of the last base of each one, a computer deduced the sequence of each strip of DNA.

When Venter gave reporters tours of the facility that summer, he liked to bring them to this floor first and remark that it housed the second-largest sequencing facility in the world. Then he'd lead them to the floor below, which buzzed with the activity of some 150 sequencers and stretched as long as a football field. "And this," he'd say, "is the largest sequencing facility in the world." The line never failed to get a chuckle.

The back of this formidable room held space for another seventy or so machines, still arriving in wooden crates. Venter's choice of artwork had already been installed: a metallic blue DNA double helix sculpture that spanned most of the ceiling's length.

The stream of data from all these devices, each one generating thousands of letters of genetic code every hour, flowed through hundreds of cables threaded through the building's hollow floors to a glass-walled enclosure housing towering blue-gray computers with blinking green lights. These computers, from Compaq Computer Corp., packed power that rivaled the monstrous machines used for modeling nuclear detonations or world weather patterns. Venter liked to claim that his computer facility could run a nation.

Celera had already sold subscriptions to its DNA database to three companies—Amgen, Pharmacia & Upjohn, and Novartis—for millions of dollars a year. These companies could remotely tap into their dedicated computers in the data center to download or search the accumulating fly sequence along with information about a large number of the fly's genes.

All this computational cargo was so precious that the data center was inaccessible to all but a select few who could pass the identity checks flanking its imposing white door: a handprint scanner and a number pad requiring a code. But usually employees did not have to enter the data center. Instead, TV cameras peering into the room sent pictures to a semicircle of black monitors situated in a separate NASA-style command and control center staffed by human beings, who also kept track of the output from the sequencing machines.

So far, the fly DNA was a collection of unassembled tiny fragments. Fueled by caffeine, Myers and his team had developed much of the software they needed to assemble the genome and had tested their program on a full-scale simulated fly genome. They were itching to use it on the real fruit fly DNA data, but there still wasn't enough of it to work with.

By the end of July, Celera's sequencers had collected five hundred million letters of the fly's genetic instructions, about one-third of the raw data the team thought they'd need to completely piece together the fruit fly genome. (Researchers were sequencing 120 million bases of the fly's genome.) And in mid-August, after being dogged in the early months by machine breakdowns, Celera's sequencing factory suddenly started producing data at a phenomenal rate. Though it had taken three and a half months to collect the first half of the fly's DNA sequence, the data center logged in the final 50 percent in just two weeks. In early September, Celera officials announced that their massive computer systems had logged in the last (1.8 billionth) letter of genetic code for the fly.

But Myers, Sutton, and the rest of their team, which by then included eight software designers, didn't wait for the last letter of code.

Myers was scheduled to present his team's first results in mid-September at the biggest sequencing conference of the year in Miami, Florida.[3] To make that deadline, he and his team had to start assembling well before September 1. In mid-August, Myers's software started massaging tens of thousands of raw snippets of sequence through the big Compaq computers. Though Myers had assembled the simulated data several times already, he was nervously waiting to see what would happen this time. Simulated data never precisely matched the real thing.

Meanwhile, the Baylor and Berkeley teams had almost finished their map of the fly's genome. They'd ordered seventeen thousand pieces of fly DNA from a library made by Pieter de Jong's team at Roswell Park Cancer Institute in Buffalo. So the backup was in place if Myers's assembly failed.

But it wouldn't be necessary. After two weeks of number crunching, Myers had his first preliminary assembly with the data Celera had accumulated as of mid-August. There were numerous holes where no DNA data had been found. And the big stretches of fly genome the program had created were not ordered or pinned to the chromosomes. But Myers was ecstatic. He had tested it against all the external data. It was dead-on. Though this was just the fly genome, Myers knew then that his assembler was going to work for the human as well.

When Myers presented his results in Miami, a friendly heckler called out: "My God, they're bloody gonna do it!" After the talk, several former skeptics told Berkeley's Gerald Rubin that they had changed their minds about Celera's strategy for the genome. They were now convinced that Myers's strategy would succeed.

But Myers and his squad still had work to do. They'd written a half million lines of computer code so quickly that it was riddled with bugs. They tinkered with the program for several weeks and then reassembled the genome using the improved program. The genome coalesced into longer stretches of consecutive pieces called scaffolds. Myers's team also fed its program the rest of the 1.8 billion base pairs of data produced by Celera's sequencing factory, which had read the fruit fly genome an average of ten times over. The additional data closed more holes in the initial assembly.

By late October, Venter, Adams, and Rubin decided it was time to begin making sense of what they had so far. They organized a novel kind of get-together, which they called an "annotation jamboree." They invited biologists and computer scientists from academic and company labs to come together to sift the sequence for genes and start to determine what those genes did. Usually, the same researchers who spelled out the sequence did this annotation. So the whole jamboree idea, Rubin said proudly at the time, "was really consorting with the enemy."

Almost nobody refused the invitation. Fly biologists were eager to get a sneak peek at the newly sequenced fly genome, which had not yet been made public.

For eleven days starting on November 8, fifty computer scientists and biologists spent sixteen hours a day holed up in cubicles at Celera's headquarters trying to make biological sense of the fly's genome. The scientists broke away from their computer screens only occasionally to discuss their findings with nearby colleagues. They ate at the sound of a gong, which indicated that take-out food had arrived.[4]

The results of the jamboree, along with the final assembled sequence, were revealed in the March 24, 2000, issue of *Science*. Together, the teams had deciphered more than 97 percent of the fly's genome, and the jamboree participants had ferreted out eleven thousand new fruit fly genes in addition to the twenty-five hundred previously known ones. Despite this huge crop of new genes, the total number of genes they estimated for the fly—13,600—was surprisingly small. The worm *Caenorhabditis elegans*, with tenfold fewer cells than the fly (a thousand versus ten thousand), has eighteen thousand genes.

To the delight of the fly people, as *Drosophila* researchers call themselves, they discovered that fly genes are very much like ours. The fly had counterparts for 177 of the 289 known human disease genes.[5]

The results published that March were just a first stab at unraveling the secrets in the fly's genome, as they were based on computer predictions that had yet to be verified in the laboratory. Still, the sequence revealed a gold mine of information and hypotheses to test, and outside scientists were pleased.[6] "It really is a grand accomplishment," said Baylor's Steve Scherer, who was part of the collaboration.

But the big question remained. Would Myers's strategy work for the human genome? It looked promising, so promising that the publicly funded genome project decided to adopt a similar approach for the mouse genome. And *Science* magazine reported when the fly paper came out that "Venter has erased most people's doubts that he will complete the human sequence later this year."[7]

But not everyone was convinced. The fruit fly genome is roughly one-twenty-fifth the size of the human genome. "I think the jury's still out" on whether shotgun sequencing will work for the human genome, Scherer opined, pointing out that the whole genome shotgun approach did not succeed for the fly totally on its own. "*Drosophila* shows that a computer does a very nice job, but it can't possibly sort it out without the guidance of a map," Scherer said.

Myers struggled to correct the misconception that maps were essential to his assembly. Myers's assembler had spit out the fruit fly genome in the form of twenty-five big stretches of sequence, each of which was at least a hundred thousand base pairs long. To correctly position these chunks of sequence onto the fly's chromosomes, Myers's team did use little STS signposts on those chromosomes— like those Lander found in the human genome—which Rubin's team had discovered. However, they did not use the more detailed maps made by Berkeley and Houston teams under their collaboration.

The backup maps did serve one very important purpose, though. Myers used them to check that his assembly was correct.[8] Without the maps, Celera would have had no way of knowing that it had an accurate fly genome. In addition to providing this critical confirmation, the Baylor and Berkeley teams closed thousands of gaps in the genome left by Myers's computer program.

Rubin and Venter's joint publication became the most cited scientific paper of 2000. Celera's sequencing of the human genome was then well under way. Though some researchers hoped that a similar arrangement might be worked out for the human genome, there was very little collaborative spirit between the Human Genome Project and Celera.

21

MEDICINE MAN

Bill Haseltine didn't care much about whether shotgun sequencing worked or not—for the fly or even the human. Indeed, he felt the entire Human Genome Project was a bit beside the point. "I call the Human Genome Project a technofolly," he said one day at his Rockville firm in spring 1999, shortly after the project decided to rev up its sequencing operations. "It's more about poetry of the human spirit and our aspirations to explore the unknown than it is about anything practical." He underlined his point, as firm as it was unpopular: "The rhetoric surrounding the Human Genome Project is almost utter nonsense. It's delusional. It's the same kind of delusion we engaged in when we put the man on the moon. I'm personally not opposed to delusion, once you know what it's about, but you should not mistake it for practicality." The information provided by the Human Genome Project could not be quickly translated into new ways to treat and cure disease, he explained.

"There is a much faster, more systematic way to create knowledge of genes for pharmaceutical application, and that's what we did," Haseltine said. He felt that his stash of genes, already stocked and accounted for, was the essential blueprint of humankind. He boasted that he had in his freezers, embedded in a wonderland of thick frost, plastic plates containing pieces of more than 95 percent of all human genes. Never mind the critics who considered his claim pure hubris. Haseltine felt he already had the genetic goods for virtually ending human disease. So while Celera scrambled to get up to speed in its race to the moon, as it were, Haseltine worked furiously to translate his genetic treasures into a medicine cabinet of compounds that might someday rebuild most, if not all, our body parts. That, at least, was his vision.

Throughout 1997 and 1998, HGS scientists had added to their store of genes for unknown proteins that they thought might be good drugs or drug targets. By late 1998, they had some fourteen thousand genes in hand and used these genetic templates to produce proteins. They then started screening the proteins on a massive scale for their effects on many cell types. By automating the process, they hoped to find functions for large numbers of genes and eventually hit on a few life-saving remedies.

Every Friday morning during the winter of 1999, skilled technicians arrived at work at 6 A.M. to bathe in a special solution some two thousand different cultures of spidery-shaped human cells, getting them ready to produce two thousand different human proteins. For three days, the cells squirted out the proteins into the surrounding soup before lab scientists scooped up the invisible product the following Monday and delivered it to company researchers for testing.

In one department, researchers used robots to test the proteins, ninety-six at a time, for their effects on cells. In under a minute, the machine revealed which proteins produced a specific chemical change in the cells by reading a fluorescent signal that appeared above the small pits containing those proteins. Some two dozen plates shuffled through the machine, spitting out a positive or negative response for each of the two thousand proteins, gleaning a small clue about what each one does in the body. The clue was then confirmed and followed up by a separate group of researchers conducting more in-depth studies. Much later, proteins with interesting properties would be mass-produced and tested in animals for therapeutic effects. The process was repeated over and over again.

By spring, two proteins and one gene from HGS's initial crop of drug candidates were showing promise as potential medicines. One was a growth factor that ordinarily springs to action in the body when

there is a wound in epithelial tissue. HGS researchers hoped this substance would benignly attract new cells and connective proteins to a
human wound, healing it. They had already seen the protein mend
straight cuts in rats similar to surgical wounds in people and close
wounds in animals that usually fail to heal because of poor blood supply
to the wound.

A second protein protected blood-forming cells in mice from
repeated rounds of toxic chemotherapy. Haseltine hoped it would have
a similar protective effect on the bone marrow of cancer patients. And
the therapeutic gene, which produced a factor that spurred the growth
of new blood vessels, caused new capillaries to sprout in the legs of rab-
bits. Haseltine wanted the gene to create new conduits for blood in
patients with severe atherosclerosis that had blocked major arteries.

Based on these animal studies, all three compounds had finally
made their way to the ultimate testing ground: humans. Haseltine was
thrilled, seeing the milestone as a realization of the plans he'd laid long
ago. "For me, this is like a butterfly emerging from a cocoon," he said.
With each molecule rested Haseltine's hopes of proving the worth of
his revolutionary gene-based approach to drug development.

By then, HGS had grown from a couple of offices to a complex of
buildings where hundreds of gene hunters embellished upon Haseltine's
atlas of gene anatomy. To keep his staff happy and focused, Haseltine
renovated and redecorated every structure his employees occupied. He
put in windows so that light beamed through the end of every hall and
along the walls just below the ceiling. He handpicked a gallery of artwork to mirror the jobs performed in every room. Circular shapes and
colored strands surrounded the biologists working with DNA and cells.
Researchers refurbishing the computerized human gene map worked
under scenes of the Mediterranean Sea, views of the shores of Europe,
global satellite maps, and images of planets, star bursts, and galaxies.

Other art served to remind the employees of the ultimate goal: to
treat and cure suffering humanity. In one painting, a human body
made from clay was in the early processes of dismemberment, its
pieces floating. In another, jumbled body parts bore labels. Teeth. Eye.
Hair. Arm. In a third, a humanoid figure had been reconstituted out of
books—information.

In the front of the lobby of the HGS headquarters building, Haseltine displayed thumb-size vials containing his star molecules, each one entombed in glass. Down a few short stairs from the reception area a miniature model depicted the manufacturing plant at which those proteins would, if all went well, be mass-produced for large-scale tests in humans and then, hopefully, for distribution as approved drugs.

The eight-thousand-square-foot manufacturing plant was nearby, its redbrick exterior sporting a long insignia of interlocking red and blue strands reminiscent of the DNA double helix. The rooms contained tubes and giant metal vats needed for production to commence. One room, designed to deliver purified air of various grades, housed little more than big, shiny, silver-colored air ducts.

Behind the lobby's curved glass walls stood a large bronze statue of the god Mercury carrying a caduceus (a winged staff with entwined snakes), a symbol of healing. Mercury served to announce the raison d'être of this $45 million factory: to produce what might be the first triplet of drugs to emerge from the human genetic gold rush.

But even if this first crop of compounds didn't make it to market, Haseltine was confident that other proteins would. By mid-1999, HGS was evaluating more than a dozen new potential human gene and protein drugs for possible clinical tests. Among them was a protein they called BLyS, for B-Lymphocyte Stimulator, that caused certain immune cells to grow and become activated in mice, and that might be used to treat ailments like AIDS in which the immune system was impaired. Numerous other researchers had been searching for this compound. But thanks to the power of HGS's genetic database, HGS researchers discovered it first, beating those who considered themselves experts in the class of molecules of which BLyS was a member.

In addition, Haseltine dramatically expanded his team's drug repertoire beyond therapeutic protein drugs to antibody-based medications. Instead of helping patients by adding to their store of a body protein as HGS's protein drugs did, antibody drugs would do the opposite—lower levels of certain proteins that were being produced in excess. In the body, these immune system proteins stick to and inactivate circulating molecules associated with disease-causing organisms.

By making antibodies against each one of the therapeutic proteins HGS had isolated, Haseltine thought his team might produce a treatment for the mirror-image disease—for instance, an ailment caused by a hyperactive immune system rather than an inactive one. Antibody medications thus opened up an entirely new category of drugs based on the original products of Haseltine's protein factory. Since HGS already had discovered specific, and secret, protein targets for these antibodies, Haseltine felt he had landed upon yet another uniquely powerful drug-development scheme.

Haseltine forged new alliances with several firms sporting state-of-the-art techniques for efficiently crafting human antibodies. In early 2000, HGS inked a $67 million deal with a British biotechnology firm called Cambridge Antibody Technology Group. HGS got equity in the company and rights to use its technology on an unlimited number of proteins to produce new therapies and diagnostic tools.

The two companies first focused on BLyS, the immune stimulant. Evidence garnered by HGS and other scientists suggested that higher-than-normal blood levels of BLyS played a crucial role in the serious autoimmune disorders lupus and rheumatoid arthritis. The data indicated that using an antibody to lower these levels might be an effective treatment. HGS and Cambridge Antibody were sorting through thousands of BLyS antibodies they'd identified to select the most promising one to introduce into humans.[1]

While the two teams were looking for a potential BLyS antibody treatment, HGS got the go-ahead from the Food and Drug Administration for the first human tests of BLyS protein itself. Since BLyS was found to stimulate the immune system to make protective antibodies, doctors would deliver it to patients with a rare disorder called common variable immunodeficiency that prevented them from making enough antibodies. Sufferers were very susceptible to infections, often causing repeated bouts of pneumonia, bronchitis, and sinusitis, among other ailments. The HGS team hoped that BLyS would restore the ability of such patients to produce their own antibodies. This would relieve them of the lifelong antibody infusions they needed to protect them against recurrent infection, a treatment with its own side effects and dangers.

In September 2000, HGS's wound-healing drug Repifermin successfully passed its first clinical hurdle. Nurses at fifteen medical centers gave the topical treatment to a total of ninety-four patients with large leg sores called venous ulcers that refused to heal because of problems in the underlying blood supply.

The results were promising, but not spectacular. In participants with the smallest wounds, the drug appeared to work quite well to partially heal patients' wounds. More than 80 percent of the patients receiving the highest drug dose saw their wounds close three-quarters or nearly all the way, compared to just 40 to 50 percent of those receiving a placebo spray. In addition, the number of patients seeing significant (but not total) healing corresponded to the dose, with somewhat more of those who got the higher dose responding to the treatment than those who got the lower dose.

Overall, just 50 to 70 percent of the patients receiving Repifermin experienced significant wound closure compared to 35 to 45 percent of those getting the placebo. But the drug showed no statistically significant ability to completely close a wound compared to a placebo. Even so, the drug appeared to be safe, and the results were good enough to pave the way for a larger trial, which HGS announced it would soon be launching.

The Washington Post reported dramatic testimonials from patients whose wounds had miraculously healed after participating in the trial.[2] One fifty-three-year-old Philadelphia man faced the possibility of having his leg amputated because of a painful, oozing sore that spanned the length of his calf. None of the standard treatments had worked. But after a nurse sprayed his leg twice a week for a month with an enigmatic potion, the skin grew over the wound and eventually closed it. He was thrilled, although, in fact, he didn't know whether he'd actually received Repifermin or a placebo. A woman who participated in the trial was so happy with the healing of her wound that she was going to write Haseltine a thank-you letter.

SmithKline Beecham regarded the compound's odds of success as good enough to bet on. It exercised its option to codevelop and then help shepherd Repifermin through large-scale clinical trials in return for a share of future profits.

Repifermin was also by then undergoing early-stage clinical trials to test its safety and effectiveness on internal wounds of the mouth, throat, and gastrointestinal tract. Haseltine hoped the drug might, for example, help heal internal ulcerations caused by cancer treatments and thereby enable cancer patients to tolerate higher doses of cancer-fighting chemicals.

As HGS's crop of therapeutic proteins wound their way through human trials, Haseltine was looking for promising technologies to make them even more effective. One problem with protein drugs, he realized, was that they were rapidly broken down in the blood, making it difficult to maintain therapeutic levels of the molecules. Typically, one would need to inject high doses of a drug or to inject the drug at frequent intervals. But high doses often worsen side effects, and frequent injections are inconvenient. Haseltine and his team were looking for better ways to deliver their proteins, and they began investigating dozens of different technologies to help them do that.

One day they came across a small biotech firm in Pennsylvania called Principia Pharmaceutical Corp. Principia researchers had developed a way to extend the life of proteins in the body by fusing them to the blood protein albumin, an abundant and stable protein that is also found in egg white. Fusing therapeutic proteins to albumin, Haseltine figured, would give the therapeutic molecules the same stability as albumin, enabling them to stick around longer in the body to treat a disease. This could make the drugs more effective, safer, and more convenient. Patients might get a shot, say, every month rather than every day.

The data showing that the technology worked as advertised was very preliminary, but Haseltine liked the concept. Within six weeks of learning about the company, HGS bought Principia for $120 million in stock and set about integrating Principia's staff and technology into HGS. HGS immediately began looking into creating long-acting versions not only of HGS's novel protein drugs but also of several of the other therapeutic proteins on the market.

∞

IN ADDITION TO keeping a watchful eye for immediate opportunities for his firm, Haseltine was pondering far more distant possibilities.

Seated on a soft paisley couch in his ornate New York City condominium one Easter Sunday, he launched into his visions of medicine's future. The genetic blueprints for the molecules that can regenerate every human organ are already on the drawing board, in the form of genes that repair skin, and grow hearts, blood vessels, and brain cells. Mankind is now, he insisted, at the dawn of a new era of regenerative medicine, a discipline that will, in a few decades, enable most, if not all, of our body parts to be rebuilt with proteins produced by our own cells.

But nobody faced with the dusk of middle age could be content with that. In Haseltine's mind, the future also held the possibility of making body parts young again to dramatically extend life. The process would entail dialing back the genetic clock on our cells to age zero, just as happens in a fertilized egg. In such a state, cells are not yet programmed to make a specific body part, such as an arm, leg, or liver. But applying Haseltine's regenerative compounds to the cells would coax them into developing into whatever body structure was desired. Haseltine likened the process to moving a pawn along a three-dimensional chessboard. So instructed, these very young cells could be used to rebuild older organs from within and, in the process, make them young again.

Haseltine seemed eager for his rejuvenation shot, something he'd probably have to wait at least two decades for. With that, his various organs and tissues would be systematically regrown into healthy, younger ones. Every ten or twenty years, he envisioned, people would get a new infusion of cells, which would slowly regenerate every body organ, including the brain, replacing what was there before. In other words, advancements in medicine would enable human beings to molt.

If this fifty-four-year-old man morphed into a younger version of himself, he'd retain the memories of his mother's illnesses, chemistry class at Berkeley, the competition to complete his thesis at Harvard, the births of his two children, the deciphering of the AIDS virus, battles with Craig Venter, and the building of his gene factory. The memories would just accumulate, as life stretched on, decade after decade. "Just as we are in pretty good shape between the ages of 15 and 25,

I hope someday in the future we'll continue to be in good shape between the ages of 250 and 265," Haseltine said.

Haseltine's dreams would undoubtedly expand, too, something difficult to imagine on a day when they seemed to become more elaborate by the minute. People could be made more durable by replacing the carbon in their bodies with silicon, he was saying. That way, we'd also be less likely to die in an accident. "I think this will actually happen," he said.

22

THE RACE

The Human Genome Project was rolling with its accelerated plan to sequence the human genome in the spring of 1999. Everyone was nervous. Though at this point Celera hadn't even begun sequencing the fruit fly, much less the human genome, the public genome teams felt its presence like a hot breath on the back of their necks. They needed rapidly to dump as much sequence as possible into the public domain to prevent Celera from seeking patents based on the sequence it collected. They were also determined to prove wrong Venter's accusations, trumpeted by the press, that their project was inefficient and slow.

Francis Collins was working hundred-hour weeks almost without break. Though the previous summer he'd married Diane Baker, the genetic counselor he'd met back in Michigan, he saw his wife only on alternate weekends. His two daughters were grown, employed as a doctor and a social worker, and had recently married as well. He figured it was a time in his life when he could do this.

When he'd signed on as genome project head, Collins no doubt had meant to make the history books. But his entry was not yet assured to read just as he'd like. Collins was still seething over a line in an article in *Time* magazine that had appeared in January—the same one featured prominently in the lobby at Celera—which read: "The federal

genome project, critics carp privately, has been shockingly misman-
aged and is sorely lacking in vision." By the end of the project, Collins
wanted to leave no doubt about his vision and management skills.

He'd started by upping Venter's challenge. By declaring that the
genome project would finish a rough genome by spring 2000, Collins
forced Venter to speed up his project by a year. To do that, Venter
decided to halve the amount of sequencing his team would do, covering the genome at fivefold instead of tenfold redundancy, and to rely
on the rapidly accumulating public data to fill in the gaps. The strategy
made good business sense, since Venter needed to create the best
genomes possible as quickly as possible for his customers. But it would
raise thorny questions about scientific credit.

Even so, Collins's team still had to make good on its promise to
sequence 90 percent of the genome in a year, after it had taken a
decade to spell out the first 15 percent. This process began with the big
chunks of human DNA that Pieter de Jong's team in Buffalo had
cloned into bacteria and sent to the sequencing centers. These were
still too big to sequence whole, so technicians at each genome center
used sound waves or enzymes to randomly break up copies of each
chunk into millions of overlapping fragments from a thousand to six
thousand base pairs long.

They then spelled out the sequences of these shorter fragments,
reading about five hundred base pairs from one or both ends of each
one. This process generated many overlapping short stretches of DNA
that computer gurus would reassemble to re-create each chunk. Eventually, the sequences of the chunks would be linked to produce the
sequence of each chromosome. Using this process, the labs were now
aiming to sequence four times as many bases as the human genome
contained, statistically plenty to nab more than 90 percent of the
genome's bases.[1]

To accomplish this required a huge boost in sequencing prowess
for most of the sequencing centers. But that wasn't the only challenge.
De Jong's huge library of four hundred thousand human DNA chunks
had been generated randomly, without respect to their place in the
human genome. So the chunks were completely out of order. In addition, many of the chunks were redundant, because de Jong had wanted

to ensure that this library included pieces from virtually every part of the genome. So now researchers faced the task of sifting through all of the chunks to select a subset of about thirty thousand that they could put in order across the genome. That would be the subset of chunks the teams would actually sequence.

The original idea was to order these chunks before sequencing even began. But the chunks weren't available until late 1997, when de Jong finished remaking the libraries to conform with ethical standards. So nobody had done this yet. The revised plan was to do the ordering and sequencing simultaneously.

A team at Washington University in St. Louis led by a young biologist named John McPherson had taken on the bulk of this additional burden, aiming to do the ordering job for the whole genome at once. In theory, teams at the other genome centers were supposed to be helping out, by ordering chunks in the region they were going to sequence. But MIT's Eric Lander refused, arguing that it was silly and inefficient to duplicate Washington University's work. Instead he asked McPherson's team, under the guidance of center director Robert Waterston, to send him ordered chunks from chromosome 17, which MIT was supposed to sequence, plus whatever other regions needed to be done. Waterston relented, although he wasn't happy about it.

Lander concentrated on sequencing, which was also the far more glamorous task. His first move was to submit a $40 million order for 125 of Perkin-Elmer's new automated sequencing machines, the 3700s. He planned to get rid of his team's entire collection of forty older sequencers and build a factory with the new ones. It was a gutsy move, especially because Lander's grant wouldn't come close to covering the cost. The silver-tongued Lander convinced the Whitehead Institute board to take a loan to buy the machines.

Nobody followed his lead. The other G5 members ordered a dozen or two of the new machines, but they refused to risk everything on them. At Washington University, Elaine Mardis's sequencing squad started cursing the three dozen 3700s they'd ordered almost as soon as they arrived. Not only did the new machines keep breaking down, but their output was initially only half what Applied Biosystems had promised. For a while at least, the Washington University team found them no more efficient, and far less reliable, than the older machines, despite costing twice as much.

If Lander was having machine problems, he wasn't admitting to them. Because he had purchased such a large number of the new machines, Applied Biosystems sent over on-site technicians to help the MIT group work out the kinks. His team was determined to make the machines work.

As the new sequencers trickled in during May, June, and July, Lander's operation gradually got up to speed. Three different factory-style conveyor belt systems transported plates full of DNA samples through a dozen different stations, where robots performed the biochemical steps needed to prepare the DNA for sequencing. One system purified the pieces of human DNA from the bacterial cells that copied them; another set up the sequencing reactions; a third pooled the samples just before sequencing. A staff member transferred the DNA samples from one robotic system to the next and pressed go. The machines did the rest.

For the coming year, Lander had secured by far the most funding for his lab: $53.4 million for FY 2000 compared to $42.6 million for St. Louis and $17.6 million for Baylor. But Lander's operation was still a far cry from what Lander had promised. The ever optimistic Lander nevertheless began hounding the St. Louis team for more ordered chunks to sequence. Waterston resisted. It made no sense to Waterston to give Lander more chunks until he had finished with those he already had. Waterston needed to keep some ordered chunks for his own more productive sequencing operation, and his team's supply was quite limited.

Lander insisted that his laboratory was anticipating a surge in sequencing speed that fall and he needed the chunks ahead of time because it took months to prepare them for sequencing. But Waterston was not confident that Lander's projections were realistic. To get proof, Lander needed something to feed his sequencing pipeline, and getting the necessary DNA from St. Louis was apparently not an option. The solution was obvious, but Lander knew it was likely to cause trouble. So he told no one what his team was going to do.

OVER IN ROCKVILLE, Celera convened an independent review board to approve its protocol for selecting DNA donors and getting their informed consent. Like de Jong in Buffalo, Celera advertised for donors in the local newspaper, but it also received a lot of unsolicited calls from people eager to participate. There was one volunteer who didn't have to call: J. Craig Venter.

The genome project's official guidelines strongly discouraged using lab staff as DNA donors, because of "the increased likelihood that confidentiality would be breached" and "the potential that the choice of persons so closely involved in the research may be interpreted as elitist." But Venter felt almost obligated to volunteer. "How could I ask other people to give us DNA if I wasn't willing to do it?" he said.

Besides Venter, Celera drew blood from twenty other volunteers. Of those twenty-one samples, Celera researchers selected DNA from two men and three women from diverse ethnic and geographic backgrounds. None of the donors supposedly knew whether he or she had been selected—except Venter, but he didn't tell anyone, including his own staff.

The plan was to sequence one of the men first, followed by a somewhat more cursory sequencing of the other four donors to yield a composite genome sequence. Working in small beaker-filled rooms around the corner from the sequencers on the fourth floor of Celera's main building, Hamilton Smith and Cindy Pfannkoch began preparing samples for sequencing in July. As with the fly, they broke the DNA into pieces of two sizes, either two thousand or ten thousand base pairs long. The plan was for Celera's sequencers to read five hundred base pairs from each end of all the pieces, providing strips of data separated by one thousand and nine thousand base pairs, respectively. Knowing the relative positions of these five-hundred-base-pair strips would provide valuable data for the assembly.

By Labor Day, Smith's DNA pieces were ready. Celera began sequencing them in its two big sequencing centers on the third and fourth floors, by then stocked with their full complement of three hundred machines.

BY THIS TIME, the publicly funded genome centers had deciphered about a quarter of the human genome, up from 17 percent in May. They were already a month behind their schedule, in part due to delays in receiving the new sequencing machines from Applied Biosystems.

The genome project's leaders felt the strain. The tight deadlines, machine breakdowns, and unfriendly competition weighed especially heavily on Waterston. He wasn't sleeping as well as he used to. "You know, we are pushed to push the goals, and these are very aggressive goals," he said.

But the pressure from Celera was the worst part. It wasn't just the threat of patents, but also the fear that their work—and that of hundreds of other publicly paid people piecing together the genome—would be drowned out by the continual blast of PR from Celera's offices. The national media had been giving Venter a huge amount of credit. In late September, *The Washington Post* featured Celera in a long story that augured success for Venter's plan to unravel the "book of life . . . faster and more cheaply than an international scientific consortium that embarked on the same task a decade ago."[2] St. Louis's Elaine Mardis lamented that her team did not have a powerful PR engine backing its efforts. "No matter how much PR they can generate, nobody's going to be able to take away from me what I contributed to this project," she said.

But the pressure did not seriously affect the technicians, a casual, cheerful group, among whom bleached blond hair, body rings, and elaborate tattoos were common. Some of them had been at their jobs seven years or longer. Team member Philip Ozersky even kept working after reaping $2.7 million from auctioning St. Louis Cardinals slugger Mark McGwire's record-shattering seventieth home run ball, which he caught at the September 27, 1998, game.

One reason was Waterston. He didn't talk down to his staff—despite being, as one staff member remarked, an internationally recognized "brain"—and he gave everyone their say about how things were run. Waterston held monthly lab meetings that included his entire staff of 350 people. They took the format of a talk show, with host Bob wielding a microphone and calling on a series of presenters.

The technicians were divided into four separate teams called Crick, Watson, Franklin, and Avery—after the four historic figures in molecular biology. The teams ushered hundreds of thousands of DNA samples through the sequencing process every week. They worked in overlapping eight-hour shifts, keeping the machines humming from five-thirty A.M. one day to three A.M. the next. Many employees also worked weekends and put off vacations in hopes of staying on schedule.

Gradually, the hard work started to pay off. By October, the St. Louis sequencers had spelled out two hundred million bases of human DNA, half of it polished, or "finished," with few gaps, and half of it in rougher form. The group had contributed about one-quarter of the total output of the sixteen publicly funded genome laboratories around the world and was on track to quadruple its output by the end of the year. That was twice Waterston's initial projection. Mardis had finally received a solution to the productivity problem. Applied Biosystems replaced the "goo," or matrix, separating the DNA fragments inside the machines' capillaries, greatly improving the 3700s' output.

However, Waterston had trouble staying motivated in a big collaborative enterprise. The G5 labs had formal telephone conferences every week and face-to-face meetings every three months, and they were all governed by ground rules. Waterston felt that human sequencing wasn't as much fun as sequencing the worm had been. "We've yoked ourselves to the common good and so are more constrained," he said.

By fall, Lander's secret plan was also becoming apparent. Instead of waiting for Waterston's group to send him the ordered chunks of DNA he felt he needed, Lander had instructed his researchers to pick random chunks from de Jong's collection and prepare those for sequencing. The sequences Lander and his team were producing from this random effort were scattered all over the genome rather than being confined to the regions they had been assigned. Lander's sequences also overlapped excessively with other sequenced chunks. This annoyed other groups, who felt that MIT was being wasteful and treading on their turf.

But Lander had to have something to feed his rapidly expanding enterprise, which had then about quadrupled its output. And since Celera was playing hardball—in Lander's mind aiming to thwart the goal of a publicly available human genome—the Whitehead team leader felt there was no time to waste. "It wasn't the most elegant way to do it," Lander recalled, "but we had to move full speed ahead."

Lander was right to hurry. Almost unbelievably, in mid-October, just six weeks after the company started decoding human DNA, Celera declared that it had sequenced more than one billion bases, or fully one-third of the human genome.

The government-funded scientists were dubious. They couldn't verify Celera's claim, since the company hadn't released its data. Collins's researchers had been right to be suspicious about Celera's data-release policy. The company was already reconsidering its initial promise to release its data quarterly, having realized that quarterly data release didn't make any sense since Gene Myers's software team wasn't going to assemble any of the data before sequencing was complete.

A month later, in November, the public genome effort reached the same milestone, depositing one billion base pairs of human sequence into the public repository, GenBank. The Human Genome Project celebrated with a bash at the National Academy of Sciences attended by Health and Human Services secretary Donna Shalala and Department of Energy secretary Bill Richardson. The NIH Genome Institute also invited journalists to the party, where the major sequencing centers received awards and all the partygoers got balloons and T-shirts sporting an image of a DNA molecule.

The genome project was delivering a message to Celera: The race was far from over.

23

HOPE, HYPE, AND ACRIMONY

Despite the lack of a clear business plan, Celera's stock price began a spectacular rise in late November, doubling in value in less than a month. Strangely, the jumps in stock price did not directly coincide with any spectacular scientific achievements from Celera. One of them did, however, parallel a well-publicized triumph from the publicly funded genome laboratories.

A consortium of four of those laboratories announced on December 1 that they had wrapped up the sequence of the first human chromosome—chromosome 22, the second smallest at fifty-three million base pairs. The achievement was published in the December 2 *Nature*. But, apparently, many investors were confused and assumed that Celera researchers had sequenced the chromosome. On the day of the announcement, Celera's stock shot up 15 percent. So even when the genome project did unveil a major triumph, its researchers were hard-pressed to get credit for it.

Amid this confusion, Lander was working behind the scenes to try to broker a truce. Lander hoped that if the public and private teams could work out a way to exchange data and coordinate their publications, the media would shift away from the unseemly school-yard brawl between them and toward more important issues such as the genome project's medical ramifications. If Venter and White were committed to releasing Celera's data on the human genome sequence, as they promised they were, Lander saw no big barriers to striking a collaborative accord. Of course, such an accord could also cement all of the genome leaders' places in history and put them in line for a Nobel Prize.

Lander secretly met Venter at a hotel in Cambridge, Massachusetts, in early fall 1999. Conversing in a private hotel room and later in the empty restaurant, the two men agreed to try to put an end to the war. They tiptoed around the idea of working together. Venter was playing along in the slim hope that Lander's desire to cooperate extended to the rest of the publicly funded researchers.

After the meeting, Lander drew up an outline for a possible Human Genome Project–Celera pact. It would include publication of a joint paper, data release by Celera after publication, and a friendly, restrained approach to public relations. He appended a list of "shared principles," such as:

- Humankind will be better served if we can find a viable way to join forces to produce a better product in a more timely fashion.
- Collaboration would offer the opportunity for joint optimization of experimental strategy and analytical methods.
- The public HGP understands that Celera Genomics will be making its consensus sequence of the human genome broadly available.
- A scientific publication that combines substantial data from both sources should, according to accepted scientific practice, be a joint publication involving authors from both groups.

Lander and Venter continued their discussions throughout the fall. Lander kept Collins informed about the talks, but he didn't tell his other colleagues about them until December. By then, Lander felt he had stuck his neck out far enough and turned the negotiating job over to four others: Collins, Waterston, NIH chief Harold Varmus, and Wellcome Trust governor Martin Bobrow. They met with Venter and his associates on Wednesday, December 29, at Celera headquarters. Lander agreed not to attend. A dominating presence, he did not want to end up pushing an arrangement with which his colleagues were not comfortable. For a collaboration to work, everyone else had to be on board.

The meeting went poorly. Both sides agreed that working together would yield a better picture of humanity's blueprint. But they agreed on little else. Celera wanted several years' of exclusive commercial distribution rights to any jointly created genome. But Collins's team insisted that the period be limited to one year at most, reasoning that the Human Genome Project would produce an equivalent product within that time. When genome project scientists said they wanted to work on finishing, or closing, gaps in the joint data set and publicly release the results, White objected. He would not permit Celera's data to be used for a competing product that was simply given away. The commercial uses of that data had to be protected. Collins and his crew shook their heads. The data had to be accessible to everyone.

At the end of the meeting, White concluded it was hopeless. Venter also regarded the public side's demands as absurd, and a nonstarter for any serious discussions.

In any case, White and Venter had little need to compromise. By January 10, 2000, Celera had sequenced enough DNA—5.3 billion base pairs or almost twofold coverage of the genome—to have covered a whopping 81 percent of it. Combined with the public data, they'd compiled DNA data spanning 90 percent of the human genome. The public consortium was lagging far behind. It hadn't yet covered half of the genome. (Two months later, it had gathered data for just two-thirds of it.)

When Collins tried to reach Venter to arrange a follow-up meeting that January, Venter did not respond. On January 22, Collins did get through to White, but the conversation got him no closer to an agreement. To White, the public team wanted him to give away Celera's product, and he simply couldn't do that. White didn't feel they had any common ground. "We've been over all this," he told Collins.

Collins kept trying to reach Venter during the following weeks, but Venter still did not answer his calls and e-mails. Irritated, Collins and the other public-sector leaders shot back with an angry letter in late February. The letter scolded Celera for its intention to publish a genome that incorporated genome project data without the consent of the researchers who generated it. It also issued a warning. "A clear conclusion seems to arise from Celera's actions on and subsequent to December 29: that there is no real interest on the part of Celera in continuing to pursue this particular collaborative model." Unless Celera responded by March 6, the genome project team would consider their negotiations dead, the letter said.

Venter and White shrugged off the letter. In their minds, the negotiations had been dead from the start. But they fell into a trap. On March 5, a Sunday, the Wellcome Trust leaked the letter, stamped "Confidential," to the press. Major news outlets across the country carried the story the next day. *The Washington Post* dubbed the conflict "open warfare," and *The New York Times* summed it up as a "clash of principles and egos."[1, 2]

But if the letter's release was intended to pressure White and Venter into giving away Celera's data, the tactic failed miserably. White felt the publicly funded scientists were trying to embarrass him and Venter, a move he considered childish and underhanded. "They wrote a letter telling us what bad guys we were and then they published it—which was beneath common decency," he recalled.

On Tuesday, Venter wrote back to Collins's team suggesting that the "one-week ultimatum" and the release of Collins's letter were deliberate attempts to extinguish whatever hope remained for their negotiations. He also claimed that Collins and his team had misstated Celera's demands for intellectual property protection. The need for extended (five-year) control over the published genome, he wrote, arose only if Celera shared its raw DNA data (as opposed to an assembled sequence) with publicly funded researchers.

However irritating, the squabble did not affect the value of Celera stock. Early 2000 was the height of the biotech boom, and Celera's stock price had soared above 300 before the shares were split 2:1 in mid-February. By early March, the price had risen almost another 100 points and investors were on edge. The price, driven up by excitement over scientific progress in genetics, was far higher than could be justified by any reasonable expectation of near-term profits. It wouldn't take much to burst investors' bubble.

The puncture came on March 14, when U.S. president Bill Clinton and British prime minister Tony Blair made a relatively bland statement in support of public release of genetic data. "To realize the full promise of [genome] research," they said, "raw fundamental data on the human genome, including the human DNA sequence and its variations, should be made freely available to scientists everywhere. Unencumbered access to this information will promote discoveries that will reduce the burden of disease, improve health around the world and enhance the quality of life for all humankind."

The statement, which had been in the works for months, was apparently meant as a pat on the back for publicly funded researchers. But since it came soon after the publicly aired Venter-Collins fracas, many investors interpreted these comments to mean that private companies would soon be compelled to freely release proprietary data. White House spokesman Joe Lockhart may have made matters worse by implying in a meeting with reporters that the president planned to restrict genetic patents, although exactly what he said is unclear.[3]

Celera shares plummeted some 35 points, or 19 percent, that day, and the decline continued in following days. The stock would never again reach the peaks achieved in early 2000, as investors remained more cautious after this sudden plunge in value. Nevertheless, Celera rebounded somewhat on April 6, when the company claimed to have "completed the sequencing phase" of one person's genome. Investors figured that Celera was done with the genome, although it was not. The company was still collecting human sequence data—from the other four individuals who had donated DNA to the project—and had not yet assembled its fragmented data.

By that time, the G5 had upped their capacity eightfold in eight months. They would soon be spewing out sequence at the rate of an entire human genome every six weeks. The DOE sequencing teams were about to finish rough drafts of their three chromosomes: 5, 16, and 29. And Lander's sequencing factory had come through with the promised twentyfold jump in sequencing speed and was now the leading producer of human DNA sequence data in the public sector. Lander's robotic assembly line was preparing more than one hundred thousand pieces of DNA for sequencing every day.

This spectacular acceleration brought the public consortium close to their sequencing goal by May 8. At that point, they'd deposited 85 percent of the promised 90 percent of the genome in GenBank. The vast majority of that sequence was totally fragmented just like Celera's data. But that same week, publicly funded researchers unveiled the second chromosome (aside from chromosome 22) that they had nicely pieced together: chromosome 21. When inherited in duplicate, this small chromosome produces Down's syndrome, an ailment of mental retardation, heart problems, and other abnormalities.

As both teams entered the home stretch of the genome race, they prepared to cross its last major hurdle.

24

THE FINAL STRETCH

Celera's apparent lead in sequencing the genome was a bit of a mirage, because it had a harder assembly task ahead of it. The end game of the genome competition hinged on who could put together the fragmented DNA data into a genome most quickly. In early 2000, Eugene Myers at Celera and his exhausted team had completed their assembly software for the fruit fly genome. But adapting that program to a genome some twenty-five times larger and riddled with other complexities would not be at all easy.

Among other problems, Myers's team had to make its fly software far less memory-intensive. Using the original version of that software for the enormous human genome was impossible because the job would eat up six hundred gigabytes of memory, far outstripping the capacity of Celera's thirty-two-gigabyte computers. Myers's team also had to improve the program's method of resolving look-alike sequences because the human genome had many more of them than the fruit fly genome. In the early months of 2000, Myers and his team revised the software and tested various iterations of it on the simulated human genomes of various sizes they'd created the previous year.

At Venter's request, Myers's team also tried to write a second assembly program—one that made full use of the public project's information—including the location, and partially assembled form, of its raw DNA data. Venter wanted this in case it created a superior genome for Celera's customers, who were paying large fees to use Celera's DNA database.

But after spending two months trying to overlap and order the public's DNA segments as they appeared in the public DNA repository, GenBank, Myers's team gave up. These public DNA segments contained too many errors to yield an acceptable result. So Myers decided to cut up the segments into small fragments 550 base pairs long, put that data along with Celera's into compartments based on the fragments' locations on the chromosomes, and then use the original assembly program to piece together each compartment.

Myers and his team still had a lot of work to do, but they were making steady progress. He was quite confident that his team would be able to assemble the human genome in a few months.

Nobody could say the same for the public side. The genome project's assembly job was theoretically much easier than Celera's. Myers's team had to assemble tens of millions of randomly generated DNA sequences. By contrast, the public centers' sequenced pieces were grouped according to larger chunks of DNA. Once each DNA chunk was assembled, the chunks could just be put in order and the genome would be done.

The Washington University team had made huge progress ordering the chunks. They had now covered 90 percent of the genome with them. But this didn't help, because nobody could assemble the chunks.

Computer scientists at each of the genome centers tried to do this using a widely available assembly program called Phrap. Phrap had linked many of the small sequenced pieces, each five hundred base pairs long, into longer segments by overlapping their common sequences. But Phrap couldn't connect all of the small pieces because there were still so many missing ones. So each chunk remained fragmented, like so much molecular confetti dumped into a bag. As late as December 1999, the public genome still existed as hundreds of thousands of partially assembled fragments, and nobody knew how to rectify the situation.

But just before Christmas, Lander phoned a computer scientist named David Haussler at the University of California, Santa Cruz, to ask for his help in finding the genes in the human genome. Haussler and his team specialized in designing software that mined long stretches of DNA—strings of A's, T's, C's, and G's—for certain sequences of letters that suggested the presence of genes. He was thrilled about the prospect of applying his skills to the human genome. But then Lander warned him that he'd like him to get started right away, while the genome was still in confetti pieces.

Haussler was taken aback at the prospect of trying to find genes among so many pieces. The genes would be so fragmented they would be difficult to recognize. Nevertheless, he agreed to try, and he recruited two of Santa Cruz's most talented programmers to help him. One was James Kent, a graduate student in molecular biology, who honed his programming skills as a designer of computer animation software for video games and short films. The other was Haussler's graduate student, David Kulp, who had developed the main gene-finding software used for the fruit fly genome.

Haussler, Kent, and Kulp explored various ways of fishing for genes in the human DNA data in the public data banks. But the more they looked at it, the more impossible it seemed to make any headway with so much jumbled DNA confetti. They needed an assembled genome. In March, Haussler took it upon himself to try to create one.

This wasn't Haussler's specialty, and it wouldn't be easy. The data was riddled with errors and was changing so fast that it was difficult to know exactly what the input for an assembly program would look like. But at this point, Haussler couldn't leave it up to other publicly funded scientists to come up with something. The difficulty of finding genes without a genome was not the only issue. Haussler worried that Celera would announce an assembled genome when the public project was still left with a bunch of GenBank records.

So Haussler persuaded the dean of engineering and the chancellor of the university to help him purchase one hundred $1,800 desktop computers for the job. He began working on the problem with a postdoc named Nick Littlestone. Kulp, who also served as vice president for a software firm called Neomorphic, was busy perfecting his gene-finding methods in anticipation of an assembled genome. Kent couldn't help out just then either because he had to study for the oral exams required for his Ph.D.

Haussler and Littlestone began working out a strategy for linking the hundreds of thousands of pieces of data and ordering them in a linear fashion. They kept Kent posted on the project via e-mail. But as the work dragged on into April, Kent became concerned that the strategy they were using could not be made to work in the time they had. Celera would almost certainly be announcing its assembled genome within the next couple of months.

❧

AS THE ASSEMBLY problem became a crisis, Francis Collins was dealing with a political crisis of his own. The media coverage of his battles with Craig Venter and the race for the genome had caught the attention of other top scientific managers in the capital. Other NIH institute directors, including the director of the National Science Foundation and the undersecretary of energy, had become concerned about the impact of this very public squabble spilling over into their scientific areas and worried it might jeopardize their funding. The situation eventually caught the eye of President Clinton, who on April 7 sent a message to science adviser Neal Lane instructing him to "Fix it . . . make these guys work together."[1]

Collins hated the idea that he might be hurting the NIH and science, if unintentionally. He worried that the human genome sequencing effort was tumbling toward an unhappy conclusion in which the public and private endeavors would produce competing announcements of victory within a few months. That, he felt, would underline the unhealthy competition and detract from the project's importance to human health. So Collins began looking for a new way to make peace with his rival.

As it happened, Collins's partner in the genome project at the DOE, Aristides Patrinos, had been offering to help ease tensions for months. Patrinos was in favor of the competition between the groups. He'd supported Venter from the start because he believed his new venture was good for American science. But this competition had turned ugly. It threatened to damage his friends' reputations as well as the genome program. Patrinos revered both Collins and Venter and worried they would look like quarreling fools in the end.

Patrinos proposed that Collins, Venter, and himself meet informally, but secretly, to explore ways to end the animosity. Patrinos thought secrecy was critical. If the negotiations flopped and nobody knew about them, the three men could just pretend they never took place. Breaking the code of secrecy could result in a press debacle.

Collins was leery of such a plan at first because he was extremely uncomfortable operating solo, without consulting the genome center directors, his own staff, or his boss, the NIH director. And he didn't at first think such a radical tactic was necessary, as he felt the other efforts to forge a compromise were likely to succeed. But as the mudslinging in the media and in the halls of Congress went from bad to worse, and all other attempts at détente fell apart, Collins felt he had to do something.

In late April, he nervously picked up the phone and dialed Patrinos. "Let's try it," he said.

Patrinos immediately called Venter. Venter was pessimistic that anything would come of the meeting and didn't trust his public-sector colleagues after they pulled the letter stunt in March. Was this just another skit for the press? he wondered. But Patrinos eventually convinced him to play along if for no other reason than as a favor to him. The three agreed to meet at Patrinos's town house in Rockville on Sunday, May 7.

That morning, *The New York Times* coincidentally ran an article describing the genome end game as "a fractious exchange of charges and countercharges." The article forecast a scenario in which each team would try to preempt the other's declaration of victory by quickly issuing separate public announcements that they'd finished their draft of the human genome.

Both Collins and Venter were visibly tense at the start of the meeting. Patrinos calmed their jitters with idle chitchat, pizza, and beer, and he deftly steered the conversation to the topic of the media. He knew both men felt the press had mistreated them, and he calculated that this was a topic on which they could find common ground. They did. At the end of the two-hour meeting, the men had decided that public reconciliation might take the form of coordinated public announcements and simultaneous publication of their manuscripts. They agreed to meet again. Both Collins and Venter were pleasantly surprised they'd made so much headway on the first day.

Back in Santa Cruz, James Kent had passed his oral exams on May 1, and he immediately started experimenting with a simpler, old-fashioned way of assembling the genome. He worked by himself and didn't discuss the idea with anyone. After three days, Kent felt confident he could make his idea work and so sent an e-mail to Haussler.

Kent told Haussler he thought it was time for plan B. He'd developed a different method of assembling the genome that he thought he could parlay into a working program in a very short time. He wanted to devote his time to that, rather than helping Haussler and Littlestone with their program. Kent was telling the senior scientist he'd write an assembler for the human genome from scratch, within a month, and didn't need anyone's help.

Haussler was a realist. His and Littlestone's strategy was unlikely to produce software that would assemble the genome quickly enough. It might take a year, and Haussler knew that was far too long. So while he didn't think Kent's chances were much better—"impossible" was how he put it—he figured if anyone could succeed, Kent could. He'd seen the bearded, soft-spoken grad student perform miracles before, although nothing at this scale. His reply was as brief as it was unambiguous: "Godspeed," he wrote.

On May 24, Kent retreated to the garage in the back of his house and got to work. On the same day, Collins and Venter met at Patrinos's for a second time.

Patrinos's children were upstairs doing their homework and banished their dad and his guests, with whom they were not particularly impressed, to the basement. There, Patrinos suggested that the two sides plan a press conference in which Venter and Collins would both describe and celebrate the completion of their draft sequences. He pushed this as the best way "to shut up those who wanted to play on the dark side," as he later put it. Separate press conferences on the same day wouldn't do, he argued, since the press would use the separate venues to goad each side into criticizing the other. Collins and Venter agreed.

Their plan, however, assumed that both sides would have presentable genomes very soon, when, in fact, the assembly of both private and public data sets still hung in the balance. At Celera, forty computer processors had started chewing on some forty million pieces of DNA data using Myers's team's assembly program to cobble them into a genome. It was the first time the program was used to assemble the real DNA data. Myers wasn't sure how well his program would work. He'd find out only after the computers had spent a full six weeks massaging the data.

In Santa Cruz, errors in the publicly available DNA data were driving Jim Kent crazy as banged out his code, typing so furiously he had to apply ice to his sore wrists every day. In some places, Phrap had linked pieces of DNA that didn't belong together. In others, Kent found mouse or bacterial DNA mixed in with human DNA, or worse, misplaced DNA from a distant location in the human genome. In addition, the information from other genetic databases that Kent was using to determine the order and orientation of the pieces was also beset with errors. As a result, information from one source would often suggest that pieces should be ordered one way while that from another source suggested a different arrangement.

Kent tried to resolve the apparent contradictions by delicately ranking the data from most to least believable and programming the software to ignore questionable data when it contradicted information that was more reliable. He also used his software to smooth out rough spots in Washington University's map of the chunks.

He worked from eight A.M. to midnight, driven by the idea that if he finished a day sooner, thousands of other biologists could start their work interpreting the genome and finding clues to cures a day sooner as well. Kent also wanted to dump as much genetic information as possible into the public domain as soon as possible to prevent the possibility of patents. Patenting genes seemed wrong and unnatural to him.

By June 2, after ten days of nonstop programming, Kent produced some twenty thousand lines of computer code and put it to work on the human genome. His program, which he named GigAssembler, put sequenced pieces from neighboring DNA chunks into groups. It then identified places where the sequences of those pieces overlapped to build still longer contiguous stretches of sequence. Then it scoured more than a dozen existing databases for clues about how to order and orient all of these pieces to create segments that could be laid out in a linear fashion, one after the other, separated by the inevitable gaps.

GigAssembler looked for matches between the 375,000 disordered bits of sequence in GenBank and sequences from known genes in the public Merck Gene Index, for example. This tied together disparate pieces that contained parts of the same gene. It also used a database of previously compiled DNA sequences from the ends of the larger chunks to help order and orient the small pieces. Knowing that these "end sequences" were separated by a given number of base pairs in the genome helped the GigAssembler connect any segments of DNA in which those strings of letters appeared.

On June 6, Kent shared a preliminary assembly with Collins and the major NIH genome centers. But there was no time to rest. The program was still full of bugs, so Kent began frantically refining its algorithm.

Back in Maryland, Venter and Collins continued their talks. They met three more times and talked for hours on the phone, choreographing the main event down to the smallest detail. They negotiated to the minute how long each person would speak and in what order. They shared their speeches and worked out answers to specific controversial questions ahead of time. In addition, both men spent a lot of time simply venting their feelings and telling stories, clearing the bad blood that had built up between them. Venter found the conversations surprisingly friendly and candid, and not at all hostile.

On June 21, they settled on Monday, June 26, as the big day. Patrinos was extremely pleased, especially in light of rumors that the White House was considering hosting the event.

In those last frenetic days, Kent and Haussler put their amended program to work on the jumbled strips of code in GenBank. On Thursday, June 22, four days before the White House ceremony, Kent had his assembled "working draft" genome. Kent's heroic efforts allowed the Human Genome Project, behind in the home stretch, to squeak out a tie with Celera, whose data had taken weeks longer to assemble.

Sunday evening, June 25, Celera's computers spit out a solution to the puzzle so many thought couldn't be solved. The Celera computer wizards had tied together an unprecedented forty million random pieces of human DNA into a genome—a tour de force unparalleled in computer science. Since Celera's DNA data was far more uniform than the public project's, Myers didn't have to deal with the contradictions Kent struggled with. However, Myers's puzzle had a hundred times more pieces, and those pieces had been in complete disarray.

At the White House the next day, both teams would trumpet their successes—for the first time, everyone hoped, in harmony.

25

A GENOME UNVEILED

At ten o'clock on Monday morning, June 26, 2000, reporters, ambassadors, politicians, and an impressive crop of scientific luminaries packed into the East Room of the White House. Frederick Sanger, the genius behind the DNA-sequencing process, hid in the crowd, as did James Watson, the Nobel laureate who, along with Francis Crick, revealed the structure of DNA nearly a half century before. Charles DeLisi, the main mind behind the Human Genome Project when it began, also stood among them.

These men of genetics history gathered at this hour, because this promised to be an historic day. It was orchestrated history, but history nonetheless, and, for most, an unprecedented invitation to the White House.

At ten-nineteen A.M., the band burst out with "Hail to the Chief," and President Bill Clinton strode into the room. "We are here to celebrate the completion of the first survey of the entire human genome," he proclaimed. "Without a doubt, this is the most important, most wondrous map ever produced by humankind." Clinton was honoring both the International Human Genome Consortium and Celera together, in one ceremony, for their contributions to this map.

"Today, we are learning the language in which God created life," he went on. "We are gaining ever more awe for the complexity, the beauty, the wonder of God's most divine and sacred gift. With this profound knowledge, humankind is on the verge of gaining immense, new power to heal."

The words and the ritual seemed to have healing powers of their own. As agreed, Francis Collins and Craig Venter stood up to generously thank and congratulate each other. Collins lauded his rival for his energy, determination, and inventiveness. "Articulate, provocative, and never complacent, he has ushered in a new way of thinking about biology," Collins said of Venter. Venter applauded Collins and his team for their tremendous effort to create their draft of the genome and acknowledged that Celera could not have done it alone. "Obviously, our achievements would not have been possible without the efforts of the thousands of scientists around the world who have gone before us in the quest to better understand life at its most basic level," he said.

The two men shook hands, beaming at each other. They posed for a moment with Aristides Patrinos between them, hands tightly clasped into a sturdy three-human chain.[1]

It was an ardent display of togetherness and an important day for both teams—an attempt to etch in the public's memory that they had both finished at the same time and had reached an amicable peace accord in the end.

It worked. The press dutifully jumped on this White House affair as a huge news event. "Genetic Code of Human Life Is Cracked by Scientists" screamed the banner headline across the front page of *The New York Times* the next day.[2] *Time* magazine's cover story that week proclaimed "The Race Is Over."[3] All the major TV networks carried the news. *The Wall Street Journal* and *Business Week* each timed a major genome article ahead of the official announcement.[4, 5]

But despite all the excitement, each party's genetic blueprint was still very much a work in progress. The public consortium hadn't yet covered 90 percent of the genome as it had planned. James Kent's assembly actually stretched over just 82 percent, and only a quarter of it, including chromosomes 21 and 22, was fairly neatly reconstituted or "finished." The genome was in about 185,000 long stretches of ordered pieces called scaffolds. And even these scaffolds were not complete. There were 230,000 gaps in this genome, most of them tiny five-hundred-base-pair pores but also a few thousand gaping chasms of about thirty-five thousand missing bases.

Celera's blueprint was similarly fragmented. Myers's team didn't have time to collect precise statistics, although Celera's genome was probably composed of much longer scaffolds than the public sequence due to the power of Myers's computer program. Nevertheless, Celera's genome was also riddled with several hundred thousand gaps. Neither blueprint was yet available for public viewing.

So while everyone was thrilled to be at the White House, some of the scientists felt a little unsettled about a party at this rather early juncture when so much work remained to be done. Though profoundly moved by the ceremony, Collins was still reeling from the funeral of his sister-in-law the afternoon before. She had died of breast cancer, a sharp reality check of how far scientists still had to go before turning the knowledge of humanity's blueprint into cures.

And just as these versions of the blueprint were ephemeral, frozen for this moment for the sake of peace and politics, so were the smiles, handshakes, and promises from both sides to coordinate their efforts. While Clinton remarked that the rival teams planned to join together to annotate, or analyze, the human genome, there were, in fact, no concrete plans to do that. Collins and Venter deliberately avoided the word "collaborate."

Both teams had, however, voiced a desire to publish together. Earlier in June, Venter and Collins, along with Patrinos and Eric Lander, had quietly met editors from *Science* magazine for dinner at a Marriott hotel in Washington. There, everyone agreed it would be ideal if *Science* published both papers side by side in the same issue. After the White House ceremony, the editors at *Science* thought the outlook was even more promising. Articles in *The New York Times* and *Time* magazine mentioned both sides' intentions to publish their papers in the same journal.

∞

NEITHER SIDE HAD much to publish yet, of course, and so everyone immediately got back to work after the ceremony. Aiming for a finished sequence in 2003, the publicly funded sequencing centers ran more human DNA through their machines. Gradually the additional sequence closed some of the gaps in their genome. In addition, separate teams at each center employed chemical tricks and other ingenious tactics directed at some of the genome's most recalcitrant gaps—those left because the bacteria cloning the DNA pieces consistently refused to copy certain DNA pieces.

Kent and David Haussler debated with Collins and the heads of the G5 sequencing centers about how soon to release humanity's rough blueprint to the public. Kent and Haussler wanted time to improve their program, but the genome barons wanted to adhere strictly to their policy of open and rapid release of their results. So in a frantic few days, Kent and Haussler scanned the genome for any really embarrassing flaws and, finding none, posted it on the Internet on July 7.

The news that the code of our species was now on view spread like wildfire through Internet chat groups and private e-mails. "Hey, did you know the human genome is online?" somebody wrote. "Who put it together?" asked another. "Who is Jim Kent?" wondered a third. Some science fiction fan calculated the number of times the sequence GATTACA—the movie title—appeared in the genome and reported back the result. So many people tapped into the Santa Cruz servers for a look that a half trillion bytes of data was snatched from the campus that day, easily eclipsing any previous record of daily Internet traffic recorded by the university.

In late July, Kent posted the first version of his genome "browser" on the University of California, Santa Cruz, Website. The software enabled researchers quickly to visit any desired region of the assembled public genome and zero in on genes there. Thousands of biologists logged on, visiting various vistas of the genome they believed housed important genes and saving themselves years of work.

Some of them were modern-day descendants of Collins-era gene hunters, to whom finding disease genes had meant years of painstaking laboratory work, experimentally plucking out parts of the genome inherited with a disease and then combing those carefully for the bits of genetic code that caused the ailment. Now, though, people looking for disease genes by their position used a few keystrokes to home in on the suspect region of the genome. Within seconds, the computer would show them five or ten gene candidates. A few weeks or months of follow-up biology would then pinpoint the likely gene. Other scientists indulged in a new kind of gene search in which they scoured the genome for sequences similar to known disease genes, a tactic designed to reveal new genes responsible for the same ailments.

In addition to accessing the free public genome, many companies and universities were paying for Celera's premium genome service, which included access to regular data updates, other animal genomes, and software to help researchers find the genes they were looking for. Drug industry users such as Amgen, Pharmacia & Upjohn, Novartis, and Pfizer paid millions of dollars per year to subscribe and, in some cases, agreed to share future revenue on any drugs derived from the use of Celera's data and software tools. Universities bought the service for $10,000 per year and up. Celera's first academic customer was Vanderbilt University, followed by Harvard, the University of Cincinnati, the University of Texas Southwestern Medical Center at Dallas, and Ohio State University.

Genome scientists from both the public and private squads began trying to make some general sense of their newly discovered genomic landscapes. On the public side, a team of forty or fifty biologists recruited by Lander pored over the long strings of A's, T's, C's, and G's and discussed their results twice a week in two-and-a-half-hour conference calls. They estimated the human genome's total size, the proportion occupied by genes, and examined the distribution of genes and repetitive junk in the genome. They developed sophisticated computer algorithms to pluck out specific genes from the morass of genetic gibberish on their computer screens with the goal of collecting the total set of genes that underlie human existence.

At Celera, the sixty-person genome annotation team focused on finding genes. The team developed a computer program named Otto that attempted to pick out and enumerate all the genes in the genome. Celera also sequenced additional strips of DNA known to contain human genes. Matching these sequences against the genome helped highlight the genes. The effort was much like what Venter had proposed doing in 1991 when he asked for a genome grant to sequence ESTs.

❧

WHILE BOTH SIDES hurriedly accumulated fodder for their land-mark scientific communiqués, Human Genome Project scientists orchestrated an extraordinary campaign to dictate the terms under which Celera could publish its genome. In May, Lander wrote a nine-page memo to *Science*'s incoming editor in chief, Donald Kennedy, and the senior *Science* editor in charge of the genome papers, Barbara Jasny, in which he outlined his thoughts about whether Celera should be allowed to place restrictions on the data supporting its paper. Lander thought *Science* should insist that Celera make its data freely available, with no limits on the amount of sequence a researcher could download for his or her own use.

When a scientist publishes his work, Lander argued, the scientist trades scientific credit in return for giving up his or her discovery to the world. The author can still patent the discovery, but he or she has to make it known, enabling anyone in the world to incrementally build upon the knowledge. That was the purpose of publication, he argued.

Jasny and Kennedy encouraged input from Lander and other mem-bers of the public project, because they hoped that doing so would ulti-mately bring both parties together in the pages of *Science*. But they didn't feel obligated to give Celera's competitors a seat at the negotiating table.

Kennedy reassured Lander in early June that he thought *Science* and Celera could come up with an agreement that satisfied the terms out-lined in his memo. The journal's policy required that contributors make all data relevant to their published papers available without charge. In Celera's case, that would be the entire human genome sequence, which would be put on a Website (as the genome's three billion chemical A's, T's, C's, and G's would fill tens of thousands of magazine pages).

Ordinarily, authors deposit supporting data for their papers in Gen-Bank, the public repository that anyone can access without restrictions. Kennedy briefly considered telling Venter he had to do the same—put his data into GenBank or an equally accessible portal. This would allow competitors to repackage and resell Celera's data, rendering at least the basic sequence financially worthless.

But since Celera's genome sequence was a critical business asset, Celera insisted on some protections on it. In the first draft of the agreement he sent to *Science* in June, Venter said he'd agree to let academics look at the basic sequence for free. But he did not want anyone to redistribute Celera's data to competitors or use it to make money without paying for a license. For that reason, he could not put Celera's data in GenBank. White would not allow it. If *Science* were to insist on that, Celera's data would not be published at all.

Neither Jasny nor Kennedy wanted to close the door on the release of Celera's huge cache of important information. They also felt that letting Celera protect the value of its data was fair. So they set about engineering a compromise. The conundrum was finding a set of terms that would enable industrial scientists to look at the data while barring them from redistributing it.

Throughout the summer of 2000, *Science* editors and Celera officials danced around that issue, passing various versions of the contract back and forth. Kennedy and Jasny were trying to find wording that would satisfy members of the public consortium. They consulted Lander, Collins, and, more extensively, relatively neutral scientists such as Gerald Rubin and David Cox.

Lander kept agitating to see a copy of the evolving contract, or at least get some hint about how the negotiations were progressing. Kennedy put him off. He was not about to share details of the ongoing negotiations with authors of a competing paper. For the same reasons, Kennedy brushed off a written offer of advice from former NIH director Harold Varmus, and a team of fifteen other scientists sympathetic to the public consortium. Kennedy saw what was going on. He had to admire their chutzpah, but he was not about to let Lander's confidants learn the details of the contract.

By September, there was still no agreement. Lander and the rest of the public camp were getting impatient. At the public consortium's semiannual get-together in Paris, its members agreed that they would not submit their paper to *Science* if the journal changed its existing policy requiring the public release of data. The obvious alternative was *Science*'s rival, the British journal *Nature*.

Lander e-mailed Kennedy with news of the vote. He added on what Kennedy considered much more restrictive terms for the agreement with Celera than Lander had asked for previously, including public announcement of the license terms and an opportunity for public comment. Kennedy told Lander that a public comment period was out of the question.

☙

IN THE FALL of 2000, both sides built refurbished genomes in preparation for publication. Myers's team downloaded its final cache of data from GenBank in September and performed its final assembly for publication. When the first version of Celera's whole-genome map was ready, Venter, Adams, and others pored over it for hours at a manuscript meeting. Venter peppered his scientists with questions about the genome. At the end of the meeting, Adams's team took away a frustratingly long list of issues to try to resolve before publication.

Kent downloaded the latest public data in early October, running the four hundred thousand pieces of sequence Phrap had created through the updated GigAssembler. About the same time, Lander sat down to pen the entire forty-thousand-word manuscript for the consortium. The group still did not know to which journal they were going to submit it. They planned to make that decision in early December at the latest.

As the negotiations dragged on into late October, Lander hounded Kennedy again for details. Kennedy finally told Lander to talk to Venter if he needed to know more. Venter felt he had nothing to hide and walked Lander through the current draft of the contract. Its terms infuriated Lander, who felt they were not at all consistent with the principles he'd espoused in his e-mails. Lander stayed up until two A.M. writing an angry e-mail to Kennedy. Its subject: "Irreconcilable Differences." He scolded Kennedy for agreeing to one set of terms with Venter while he was assuring Lander of a totally different set. He was upset at being left in the dark, and having to learn the details from Venter, of all people.

Kennedy kept trying for a compromise. In early November, he arranged a conference call with advisers to discuss what he thought might be perhaps the most serious problem in the Celera contract. Celera had proposed to make its genome data available for free to academics but to charge a fee for commercial users. Asking for payment was clearly not free access. But no solution emerged until one of Kennedy's advisers, a molecular biologist from Johns Hopkins University named Bert Vogelstein, e-mailed Kennedy one Sunday after the conference call. Vogelstein proposed that Celera give industry users free access if they signed an agreement not to commercialize their results or redistribute the sequence.

At the end of November, *Science* sent the penultimate agreement to Collins and his colleagues. The terms were as follows: Celera would make its entire sequence available on its Website. Academic users could access and search it, download pieces up to one million bases long, publish their results, and apply for patents on them. If academic users wanted to download segments longer than one million bases (about .03 percent of the code), they would have to sign an agreement saying they would not redistribute the data. Commercial users got free access only after agreeing in writing not to use the results for money-making ventures. To make sure Celera kept its word, *Science* would keep a copy of Celera's database in escrow so that the journal could release it in the event of any significant violations.

Celera shipped off its magnum opus to *Science* on December 5. The publicly funded researchers were to meet in two days to decide where to submit their paper. On December 6, the emotions that had been simmering behind the scenes finally erupted in public. Michael Ashburner of Cambridge University, a geneticist with ties to the European Bioinformatics Institute, sent an e-mail to *Science*'s board of reviewing editors and the press saying he was "outraged and angry" that *Science* wasn't forcing Celera to deposit its data in GenBank. If Celera did not put its sequence in GenBank, he argued, the genome would be fractured and spread across many sites, making it much harder to search.[6] By then, however, that issue was moot. It would be nice to have all the data in one place, Kennedy admitted, but this just wasn't practical or a reasonable requirement of Celera.

The next day, the leaders of the public project met to decide where to submit their genome paper. The outcome was a foregone conclusion.

Though Patrinos still wanted to publish in *Science*, he was outnumbered by colleagues who resented Celera and couldn't stand the idea of any restrictions on how the genome data could be used. In protest, the genome project leaders decided to submit their paper to *Nature*.

In retrospect, Kennedy realized he was naïve to think the public camp could ever have been satisfied with any terms that would enable Celera to publish. He concluded that anti-Celera sentiment won out in the end. But he did not regret trying to devise a compromise. The final agreement, he knew, was better for it.

Nature and *Science* arranged to publish both papers at the same time to come as close as possible to fulfilling the expectations set in June. Both papers were announced in a February 12, 2001, joint press conference orchestrated once again by Patrinos. Celera's paper—which appeared in the February 16, 2001, *Science*—was forty-eight pages long. It took one entire page just to list all 274 authors. The public consortium's main article stretched over a whopping sixty-two pages in the February 15 *Nature*, which included companion pieces as well. Kennedy realized that if he had published both articles in *Science*, the magazine would have been heavier than the Christmas issue of *Vogue*.

26

JOCKEYING FOR THE PRIZE

The genetic blueprint of the man from Buffalo, whose DNA Pieter de Jong's team had deftly prepared for sequencing, made up about three-quarters of the public human genome sequence. DNA from another male donor prepared earlier by de Jong's group constituted an additional 9 percent of the public genome. So while newspapers widely reported that the genome was derived from the DNA of a dozen people, in fact nearly 85 percent of the data came from just two individuals. It was hardly the genetic mosaic that genome project leaders had envisioned.

But de Jong was stunned when he picked up the main genome article in *Nature* and discovered his name was nowhere to be found among its 249 authors. It was an incredible oversight, a result of both the loose organizational structure of the genome project as well as the lack of attention to, and regard for, the job of developing the essential DNA fodder for sequencing. This saddened de Jong. *Nature* published a correction five months later, providing modest consolation.

At first glance, the public and private genomes were not astoundingly different. The public team's 2.7 billion base pairs of assembled sequence covered an estimated 88 percent of the human genome. It was still a patchwork of eighty-eight thousand segments stretched out along the twenty-four unique human chromosomes. There were 150,000 gaps, the vast majority of them quite small. One-third of the genome was in finished form, up from a quarter in June.

Celera's whole genome assembly contained more pieces overall than the public's—120,000—and 220,000 small gaps. But most of the puzzle pieces were much bigger—about three million base pairs wide versus 275,000 for the public sequence. Celera's whole genome assembly also covered very slightly more of the genome than the public consortium's.[1]

On the public side, Lander's genome analysis team had picked out some of the features of the human genome's vast and intricate chemical terrain as they'd mapped it. They estimated that the genome spanned 3.16 billion base pairs, and they saw that the genes were not uniformly distributed. They found "urban" areas densely populated with genes, "rural" areas with sparsely scattered genes, and large gene deserts completely devoid of genes. Where the genes refused to live, repetitive DNA with no known function flourished.[2]

Textbooks and scholars had long bandied about one nice, round, pleasingly hefty figure for the number of human genes: one hundred thousand. Harvard's Walter Gilbert had come up with that count in the mid-1980s by dividing the size of the human genome—three billion base pairs—by his estimate for the entire span of a typical gene, thirty thousand base pairs.[3] Ever since then, scientists had used more sophisticated analyses to come up with widely varying estimates for the number of human genes.

In 2000, scientists placed $1 bets at a meeting at Cold Spring Harbor Laboratory in a game to guess the number of human genes. The winner was to be announced in 2003, the year when the complete genome was expected to be ready. The bets were all over the place, but the mean guess was about 63,000 genes. A TIGR scientist wagered that there were about 120,000 genes, similar to Incyte's count of 140,000. Francis Collins weighed in at 48,000 genes.[4]

When both Celera and the Human Genome Project counted the genes in their draft genomes, the overall number of genes was suspiciously low. Both squads used sophisticated computer algorithms to sniff out signs of genes from telltale arrangements of nucleotides that tend to appear at the beginnings and ends of genes. They also looked for matches between the genome sequence and known genes, partial genes, and proteins listed in outside databases. Both came up with about the same stunningly low estimate. It was nothing like one hundred thousand. The public consortium calculated there were thirty thousand to thirty-five thousand human genes, and Celera's figure hovered between twenty-six thousand and forty thousand. Both Celera and the consortium concluded that genes comprise little more than 1 percent of the genome's chemical code—far less than the 3 to 5 percent scientists had estimated previously.

Much ado was made in the media about how humans, with their hundred trillion cells and brilliant complex minds, could possibly be built from so few genes. *The New York Times* called it "a humiliating deflation [of] human dignity."[5] "There was almost panic because the genes weren't there," Venter quipped to the same newspaper, recalling his team's initial reaction.[6] Thirty thousand genes was not even twice the number found in Waterston and Sulston's minute roundworm, which has eighteen thousand genes and just one thousand cells. Even the tiny fruit fly has more than thirteen thousand genes.

To ease the crisis of confidence that the new gene count provoked, experts ascribed more ingenuity to human genes. It had long been known that enzymes in human cells act as little knives to whittle the initial transcript made from a gene into any of several configurations to produce any of several proteins. Because of this creative sculpting, some researchers concluded that each human gene could serve as a template for three different proteins on average instead of just a single protein. And, of course, simply adding parts to a machine is not the only way to increase that machine's complexity. One can also combine those parts in different ways. Only a moderate increase in the number of parts can result in a dramatic rise in the number of possible combinations of those parts. In this way, a vastly more complex biological machine, such as a human being, could evolve from just a moderate number of additional proteins.

The other possibility was that the computers were just wrong about the number of genes. Celera's and the Human Genome Project's quick computational approaches were, in fact, quite error-prone. They missed some real genes and identified extra genes that were either not genes at all or just parts of them. The numbers were so shaky that Celera authors dubbed their gene list "a preliminary catalog," and the consortium called its count "a valuable starting point." Both teams recognized how difficult it was to locate genes in a genome, given that genes are scarce, broken up, and had to be found in a draft genome sequence that was itself discontinuous.

Bill Haseltine, for one, was not swayed by the lowball estimates. He stood by his estimate of at least a hundred thousand human genes, and he scoffed that genome sequencing was a terrible way of finding genes. Just a few months after the papers were released, the gene count started going up. Scientists from Novartis Research Foundation compared Celera's genome with the public one and found that only fifteen thousand were common to both. They concluded that both sides had missed at least ten thousand genes, leading the Novartis researchers to estimate that the genome contained at least forty thousand genes. A few months after that, other scientists from Cold Spring Harbor Laboratory on Long Island devised a sophisticated software strategy to mine the sequence for genes that suggested a total of sixty thousand genes.[7]

The debate about whose genome was better raged on for months, as researchers on both sides jockeyed for a possible genome Nobel Prize. Soon after the papers made their simultaneous debut, Lander identified a weakness in Celera's paper and exploited it vigorously. He pointed out to any journalist who would listen that, in its report, Celera had only done assemblies that included both its own data and the public's data; it never tried to assemble a genome based on its own data alone. He concluded that Celera had never proved that whole-genome shotgunning worked. The public data had provided Celera with enough positional information to put together the genome's puzzle pieces even without the sophisticated algorithms on which Myers's team worked so long and hard, Lander contended. In short, Lander's somewhat implausible argument was that Celera never sequenced the genome at all.

It was true that Myers's team had not, at the time of the *Science* publication, assembled only Celera's own data to produce an independent genome sequence. Back in 1999, Celera had, after all, scrapped its original plan to accumulate all of the necessary data itself. In a premeditated effort to finish more quickly, Celera decided to supplement its own sequence data with the rapidly accumulating publicly available data. Celera did not deny this fact, but it was remarkably deft at obscuring it. Nevertheless, Myers had, in fact, wanted to do a separate assembly with just Celera's data before publication, but he didn't have time.

However, Celera did produce about the same amount of independent sequence data as the public consortium did at the time of the *Science* and *Nature* papers. And Myers vehemently denies that including public data in his assemblies invalidated his test of the whole-genome shotgun method. Myers contended that Lander's case didn't hold water, because it depended on inaccurate assumptions about the state of the public data. It is very hard to believe that Myers would have failed to conduct a sound test of his method.

In March 2002, Lander, along with Waterston and Sulston, spelled out the attack on Celera in *The Proceedings of the National Academy of Sciences*, a move that prompted articles in major media outlets such as the *Associated Press* and *The Wall Street Journal*.[8, 9] By then, however, Myers had assembled just Celera's data by itself and had discovered to his surprise that this yielded an even better result than merging public and private data had.

Myers's new assembly of Celera's data covered slightly less of the genome than the published assembly did—2.78 billion base pairs instead of 2.85 billion. However, it was in a mere 6,500 big pieces compared to 120,000 in the published assembly in *Science*. It also had just 150,000 gaps compared to 220,000 gaps in the *Science* assembly. And almost 99 percent of Myers's Celera-only assembly was in big "scaffolds" of linked pieces that stretched one hundred thousand base pairs or longer, with the size of the typical scaffold a whopping twenty million base pairs.

At first Myers was confused about why this assembly was so good. It had never occurred to him that he'd get a better genome with less data. But then he figured it out: There were so many errors in the public rough draft data that they interfered with the assembly process and impeded the lengthening of scaffolds like an impurity blocking the growth of a crystal. And when he thought about it, Myers was amazed by how little data was actually required—just fivefold coverage of the human genome—for an excellent assembly. The Celera team also assembled data that covered the mouse genome about five times over, and that assembly was phenomenal, too. This left little doubt that the whole-genome shotgun method had succeeded for very large genomes.

Keeping up his side of the quarrel, Myers emphasized the large number of errors he had detected in the public's draft sequence. And Venter claimed that the public consortium never assembled a true sequence from its data. "The Celera Website is the only place with a sequence of the human genome," he said. There was little basis for Venter's charge, and the errors in the public sequence, numerous as they might have been, would be ironed out in the coming years. The publicly funded genome centers were painstakingly polishing their genome, a job Celera had no intention of doing.

In fact, work on the genome was hardly done in 2001. Lander called it "half time." Collins said the draft genome sequence marked "the end of the beginning." In a paper in *Genome Research*, two gene scientists named Eric Green and Aravinda Chakravarti decided the field had arrived at "Base Camp" en route to the summit of a mountain.[10] In the new century, scientists would polish the sequence, get a more solid sense of how many genes there were, and figure out what all of those genes and their corresponding proteins did inside a cell. As the Celera team acknowledged in its paper: "This assembly of the human genome sequence is but a first, hesitant step on a long and exciting journey toward understanding the role of the genome in human biology."

That journey had just begun, as scattered biologists around the world scampered to exploit the new genetic blueprints. The process of finding disease genes by the method Collins had coined "positional cloning" was shrinking from years to mere months. Biologists had already used the public project's draft sequence to find at least thirty new disease genes this way. Among the findings: New genes responsible for color blindness and early-onset Alzheimer's disease. And with Kent's browser getting some thirty-five thousand hits a day, many others were sure to follow rapidly.

Celera's services were also spawning innumerable discoveries. Scientists at the biotech firm Immunex Corp. in Seattle, who had subscribed to Celera's services in June, were thrilled to have found new members of an important family of genes they had been studying for years. They hoped the new genes they uncovered would provide important clues to curing conditions from heart disease to multiple sclerosis. Pharmaceutical giants such as Pfizer and American Home Products thought Celera's database gave them an edge over those who were relying on just the public data. Among other advantages, Celera's software tools highlighted regions of the genome that had gone relatively unexplored. All this could mean speedier progress in producing much needed medications.[11]

At the time, less than five hundred molecules served as targets for virtually all medicines on the market. The expanding compendium of human genes promised to bring that number into the thousands, even if far fewer than half of all human proteins were suitable drug targets.

❧

THOUGH THE HUMAN Genome Project was far from over, it had finally achieved its penultimate goal, having produced the vast majority of the sequence of the human genome. For that, it could be proud. But concrete progress sketching out humanity's blueprint had come mostly at the end of a long saga of tangential triumphs and missed opportunities.

The Human Genome Project had its heroes. Charles DeLisi from the Department of Energy had the verve and guts to boost this foray into big biology off the drawing board. Melvin Simon, Hiroaki Shizuya, and Pieter de Jong invented and built the DNA fodder essential for sequencing to begin. Bob Waterston and John Sulston pioneered the earliest assembly-line sequencing efforts and voiced the first bold plans to apply them to the human genome.

In 1998, Francis Collins responded promptly and forcefully to Venter's challenge, consolidating and pushing his sequencing teams to get the job done quickly. Eric Lander came through in the final stretch, applying his exceptional talents for automating biology to sequencing and churning out the most human draft sequence of any laboratory in the public effort. And then James Kent assembled it all.

But DeLisi's goals were largely ignored during the early years after the NIH yanked the project out of the DOE's hands. James Watson, in failing to support Venter's early gene-finding invention, effectively forced Venter into the private sector and necessitated a last-minute bailout from Merck. The potential of Simon and Shizuya's BACs to provide the essential fodder for large-scale sequencing went unrecognized for years. Sulston and Waterston's early ambitions for sequencing the human genome were shouted down by the voices of caution. And James Weber's bold vision to decode the genome shotgun style suffered the same fate.

An extraordinarily gifted communicator, Collins helped the public understand what the project was all about and successfully defended its goals in the halls of Congress. He did more than anyone to ensure that there was some sequence out there that anyone could use without restrictions. But as an insider, Collins was deeply committed to the democratic NIH culture. As a consequence, he was constantly tugged back and forth by the conflicting opinions of the powerful scientists he led and, until the end, was unwilling or unable to make risky or unpopular decisions even when such decisions had a good chance of speeding progress.

Even if Collins had had the resolve, it is not clear how he could have overcome the structure of the NIH to make it more conducive to big biological endeavors like the one he led. The NIH approach had worked beautifully for small studies of a single gene or a single protein that had long been the main approach to biology. But the Human Genome Project signaled a new era for biology. It had become more like physics, where some projects required big money for centralized, large-scale efforts.

One is left to wonder what might have happened if the DOE had run the project from the start, as DeLisi's DOE colleague, Mortimer Mendelsohn, had recommended back in 1987. The project's ending date would probably have remained as DeLisi set it: 2001, not 2005. It would have focused on the efforts that were directly necessary for genome sequencing, the original goal, rather than sprinkling small amounts of money far and wide to support as many gene researchers as possible. When technology got up to speed, the DOE wouldn't have hesitated to charge one or a few facilities with the sequencing task, providing important economies of scale. It would probably have begun massive sequencing in 1995 as Waterston and Sulston had recommended.

Almost certainly, the DOE scientists, because of their engineering bent, would have appreciated and funded Venter's technology-oriented vision for decoding the genome in the early 1990s. Instead of becoming an outcast, Venter might have ended up leading the project.

As it was, Venter ended up on his own, having been far too much of a revolutionary to fit into the NIH's go-along-to-get-along culture. The same recalcitrance that cast him into the brig during Vietnam catapulted him to prominence after he left the confines of the government. Thankfully, he lived in a country that provided another productive place to which he could go.

With funding from Human Genome Sciences—but little support from his academic colleagues—Venter parlayed his EST invention into an essential storehouse of genetic information to which university researchers later angrily clamored for access. Overcoming skepticism, he and his colleagues went on to apply the untested whole-genome shotgun method to the bacterium *Haemophilus influenzae,* and thus became the first to unveil the complete genetic code of a living thing. Brushing off an even bigger groundswell of criticism, Venter assembled a team to dramatically scale up that untested sequencing method to decode the fruit fly's genetic blueprint and finally our own.

Venter's fearlessness, energy, and enthusiasm brought the human genome sequence to the world much sooner. He showed what concentrating energies and resources could accomplish. While Venter had the benefit of learning from the successes—and mistakes—of the many genome project scientists who came before him, he undeniably lit a fire under the project's leadership, forcing a much needed focusing of efforts.

Prickly, rebellious, and impatient, Venter never comfortably fit into the scientific establishment. But this ultimate outsider still had a shot at a Nobel Prize.

EPILOGUE

Venter had little time to ruminate about Nobel Prizes. After publishing his landmark genome, he had to move on. Celera's great service, a genome produced at no cost to taxpayers, was a product of a giddy time in which people were willing to invest in sexy science projects even with little prospect of profits or even significant revenues anytime soon. Celera's information business was generating $100 million per year by late 2001, but that wasn't nearly enough to cover its considerable research and administrative expenses. After the dot.com bubble burst in late 2000 and many of those new businesses went under, investors were no longer keen on data dealers. They wanted more concrete products—drugs, devices, or new diagnostic tools.

Luckily, Celera had new money, $1 billion of it, earned in a new stock offering completed in March 2000. With that, the firm plunged into the drug discovery business.

Venter's team moved out a bunch of sequencers to make room for a new laboratory geared toward proteomics, the large-scale study of the body's collection of proteins. They spread their novel suite of instruments—glitzy protein-probing devices developed by Mike Hunkapiller's team at Applied Biosystems—and staff over one entire floor of Celera. They were out to find functions for all the genes they'd identified and to determine their roles in human disease. Instead of selling that knowledge to others, Celera planned to patent it and parlay it into treatments. By studying protein interactions in the body, the company hoped to develop drugs that therapeutically disrupted those interactions that led to disease.[1]

Celera also planned to use its proteomics factory to find proteins that might serve as diagnostic markers for cancer and other diseases. It lured Kathy Ordoñez from the pharmaceutical firm F. Hoffmann-La Roche to head a new diagnostics unit in California.

Initially, Celera focused on cancer. Using the new machinery, Celera scientists analyzed thousands of different proteins from cancerous tissues and identified the ones that appeared to be particular to cancer and uncommon in healthy tissue. After selecting a number of suspect proteins, they looked up the corresponding genes in their database to find clues to the proteins' jobs in the body. Then they tried to narrow down their list of suspects to those most likely to play key roles in cancer.[2] Those would be their drug targets and diagnostic flags.

In November 2001, Celera expanded its drug-development capabilities when it bought Axys Pharmaceuticals, a small medicinal chemistry company in South San Francisco, California. Axys filled a critical gap in Celera's scientific know-how, providing expertise in crafting small molecule drugs that would hit Celera's protein targets.

But no concrete results were on the horizon. Investors wanted drugs, or at least drug candidates, and Celera had neither. All Celera had were grand plans, and that wasn't enough anymore. It didn't help that Venter himself had little or no expertise in drug development. And despite embracing other risks, Venter had long shunned the risky route of developing drugs, dubbing it "clinical trial roulette" two years before. That was Bill Haseltine's game, in which there was, Venter had sneered, a mere one in twenty chance of survival, as the vast majority of drugs that enter human trials fail.

Venter did not argue about the need for Celera to play drug roulette now. But his boss, Tony White, began to wonder whether the new Celera was still the place for the king of gene sequencing. His opinion was undoubtedly influenced by the sliding of the firm's stock price during the economic downturn of 2001. By winter 2002, the price stood at just $20 per share, plummeting from two years before, when shares of its stock sold for well over $200. Celera's market capitalization hovered around $1.5 billion. White was also still rankled by Venter's need for personal publicity. He still thought that the excessive media hoopla wasn't healthy for the business.

On January 22, 2002, Venter stepped down as president of Celera Genomics. "A realistic assessment of Craig's background is that neither he nor I have ever developed a drug," White told *The New York Times* to explain this turn of events.[3] Venter's departure was perhaps the most vivid sign to date of the changed biotechnology scene, in which selling genetic data had proven to be far less profitable than investors and biotech executives had expected.

White temporarily led Celera while he searched for a president better suited to running the transformed business. Three months later, the chief of the recently renamed Applera Corp. appointed Ordoñez to succeed Venter as president of Celera. Venter still served in the part-time post of chairman of Celera's scientific advisory board.

Meanwhile, the stock of Human Genome Sciences, the firm led by Venter's rival Haseltine, had a market capitalization of almost $3 billion in early 2002. The previous year, its share price generally hovered between $40 and $60, retaining a significant percentage of its value from its peak of $100 per share (adjusted for a stock split) in February 2000.

Haseltine's company had a big head start in the quest for gene-based cures, but it was approaching its moment of truth. Would the firm live up to Haseltine's hype? Thus far, it was hard to say. In spring 2001, the HGS team began enrolling patients in a large-scale, seven-hundred-patient trial to test the effectiveness of the wound-healing compound Repifermin to treat large, difficult-to-treat leg sores. Still under way were early tests of two other uses of this substance, those involving wounds of mucosal tissue inside the mouth and gastrointestinal tract, and of the ability of a protein drug called Mirostipen to protect the bone marrow from the often-lethal effects of cancer chemotherapy.[4, 5]

That winter, HGS began human tests of its newly acquired albumin fusion technology, which it hoped would improve the ability of an existing drug—an immune system booster called alpha interferon—to treat hepatitis C, a viral liver disease. The new drug, dubbed Albuferon, was designed to extend the life of interferon in the body to weeks from hours, dramatically decreasing the frequency of shots patients must receive. It also might enable patients to receive much lower doses of the medication, diminishing side effects such as severe fever and chills.[6] Haseltine was already plotting to produce long-acting versions of beta interferon for multiple sclerosis and of the blood-cell booster erythropoietin for cancer and kidney dialysis patients, among others.

Meanwhile, HGS begun mass-producing its first antibody drug—an antibody to the immune system booster BLyS—in its new antibody-manufacturing facility. By inactivating BLyS, the antibody was expected to help patients whose immune systems had gone into overdrive. In late 2001, doctors began giving this antibody, dubbed LymphoStat-B, to patients with the autoimmune disease lupus. It was HGS's sixth drug to enter clinical trials, and the first of many dragons Haseltine's therapeutic antibodies might fight in the future.

Haseltine projected that his group would have more than twenty drugs in clinical trials within three years. Planning for the onslaught of new compounds, he bought an even bigger, 240,000-square-foot manufacturing complex in Rockville.

At the same time, the gene-based methods pioneered by HGS were bearing fruit in the pharmaceutical industry. In 2001, Glaxo-SmithKline described results of an early human trial showing the promise of a new compound for treating or preventing cardiovascular disease that they had discovered based on information in HGS's gene database. The compound, code-named SB-435495, worked by inhibiting the action of an enzyme implicated in clogged arteries. It was the first genomics-based small molecule drug to enter clinical trials.

∞

OVER IN ICELAND, Decode Genetics had become a publicly traded company. Kári Stefánsson gave an exhausting ninety presentations in nine cities during an eight-day international road show, raising $200 million for the firm in July 2000. By then, the team had mapped a gene affecting the risk of stroke, the first one directly linked to stroke and not to risk factors such as high blood pressure or diabetes. Decode scientists had also localized five genes for various types of osteoarthritis as well as a gene for the skin disease psoriasis.

In August 2000, Stefánsson's gene-mapping teams tracked down the chromosomal home of a gene for Alzheimer's disease. That fall, they did the same for genes underlying osteoporosis and clogged peripheral arteries, and pinpointed a specific gene for the condition afflicting Stefánsson's brother, schizophrenia. Decode's scientific success had surpassed the expectations of Jonathan Knowles, research chief at F. Hoffmann-La Roche, Decode's pharmaceutical partner. In tracking down the genes behind disease, the company, he felt, was redefining disease as we know it and, at the same time, the future of medicine.

But as successful as Decode was at accumulating such data, Stefánsson also faced pressure to come up with drugs. He built a department within his firm to screen possible new medications against the proteins encoded by the newly discovered genes. By mid-2001, this eighty-person team was homing in on potential treatments for schizophrenia and psoriasis. Roche was also seeking compounds that disrupted proteins made by three disease genes Decode had discovered under their alliance—those involved in arterial disease, stroke, and schizophrenia.

At the end of the year, Decode's gene-hunting effort had expanded to cover fifty conditions, including less serious ones such as dyslexia and Tourette's syndrome. The firm was outgrowing its current headquarters and planned to move into a larger research and office complex that was under construction on land next to the University of Iceland.

Decode's stock, which was trading in the high 20s the summer week it debuted, fell to just $8 or $9 per share. The opposition group Mannvernd, then led by psychiatrist Pétur Hauksson, pressed a lawsuit against the Icelandic government challenging the law permitting the creation of Stefánsson's long-dreamed-of database. The suit was on behalf of a young woman who wanted to prevent her deceased father's records from being entered into the database.

But Decode had already built a suite of interlocking data sets and software for mining the data that Stefánsson called the "alpha version" of his database. (Its official name was Clinical Genome Miner.) As of late 2001, the database contained diagnoses on fifty thousand people and genetic information for forty-seven thousand of them. It also included the genealogy of a nation, the sequence of the human genome, and software tools for plucking out the genes and handling various other data. The only piece missing was the hotly contested mass of health care data on all patients in Iceland that Stefánsson had officially received a license to use the previous year. Decode's chief planned to begin loading that information into his system by 2003.

In quieter moments, Stefánsson often worried that, despite his success, he may have paid too little attention to his family in favor of his work, mistaking the little things in life for the big ones. But in his kind of work, big and little things often blend. Inscribed in our DNA is the drama of life, a drama that is as tantalizing as it is elusive. To transmit its text reliably from one generation to the next, DNA must be stable. But for living things to evolve in a changing environment, our genetic code must also be vulnerable to bruises and corrections. And some of those alterations will, by chance, lead to disease.

In Stefánsson's eyes, disease is the price we pay for evolution. He believes disease is necessary for life to exist. While this magnetic Icelandic scientist may not succeed in extending life through his single-minded study of our genes, he will at least have made some sense of it.

∞

AS STEFÁNSSON'S COMPANY, like Celera and others, focused less on basic science and more on pharmaceutical applications of biology, the real burden for making basic sense of our genes rested more than ever on the shoulders of government-sponsored scientists. In early 2002, Collins felt ready for this exciting new challenge. He had no intention of stepping down as the genome institute's top gun, and he was looking forward to shepherding progress that would make a direct difference in medicine. He was particularly excited about a new type of genome map that displayed points of human variation as a tool for sorting out the genetics of common diseases like diabetes, heart disease, and mental illness. He was also bursting with enthusiasm for academic efforts to find functions for the body's collection of proteins and to compare one species' genome with that of another. Now, he thought, comes the really interesting part.

But making headway up this alpine terrain will require more than just small-scale studies of important genes and proteins. It will also require bigger, more daring leaps in gene research. Biologists and computer scientists need expensive new supercomputer clusters to help them extract meaning from the human genetic code. They also need factories in which the latest machines and DNA chips help glean clues to the functions of thousands of proteins in parallel.

In the coming years, the government's genome program is likely to face less of a threat from private companies who tread on their basic science turf and threaten to tie up fundamental genetic information in commercial shackles. But it may well be racing other nations such as China and Japan to climb from base camp to the summit of this new mountain in biology.

In October 2001, Japan pledged to spend $570 million on post-genome research in 2002, more than the U.S. genome institute expected to spend in that year. With those funds, Japanese scientists planned to determine the structure and function of three thousand proteins by 2006. The effort involved the creation of seven new laboratories for analyzing protein structures and the use of the world's most powerful protein-analyzing facility for handling very large and complex proteins.[7] At the same time, China's Beijing Genomics Institute was training a supercomputer and more than a hundred of the latest robotic sequencing machines on the pig genome, which could serve as a springboard for new animal models of human disease.[8] Neither country wanted to be left behind in the next era of genome science.

To keep pace, the U.S. government will have to devise new ways to effectively support larger-scale biology endeavors. It remains to be seen whether the wake-up calls from Venter over the past decade have prodded the NIH into thinking boldly enough.

⚬❧

AFTER RESIGNING FROM the top job at Celera, Venter went sailing on his yacht in the Caribbean. Some weeks after he returned, his face popped up again on the front page of *The New York Times*. This time, Venter was claiming that he not only led the sequencing of the human genome, but also that the genome his team sequenced was mostly his own. With this astounding declaration, Venter took his penchant for opening up his life to public scrutiny to a whole new level, as now Celera's subscribers could learn Venter's own personal genetic secrets. Venter initially revealed a teaser: His cells, he said, harbor a genetic quirk associated with abnormal metabolism of fats and an increased risk of Alzheimer's disease.[9]

But whatever typos Venter discovers in his own genetic script will probably only spur him to accomplish more before risk becomes reality. So as Venter surfaced to unveil his genome encore in spring 2002, the world was waiting for him to catch his next huge wave.

Notes

In researching this story, I sometimes ran into conflicting accounts of particular episodes. To resolve contradictions, I prioritized information I could verify with an additional source or that was more consistent with other information I had collected. To compensate for fading memories, I often gave more credence to statements from earlier interviews—my interviews for the book were conducted over three years—or to articles written at the time of the events in question, even when those came from a secondary source. Although I am accustomed to prioritizing firsthand information, I quickly realized in writing the book that passing time corroded the truth far more than did the filter of an intermediary scribe.

I do not generally footnote sources for information I gleaned from face-to-face and telephone interviews or from e-mail correspondence. To do so would have been cumbersome, as such communications supplied the bulk of the book's facts and perspectives. I reference conversations only when they involved sensitive matters that I felt would benefit from further documentation.

Prologue: Unfinished Business

1. Eliot Marshall, "A Showdown over Gene Fragments," *Science*, October 14, 1994, pp. 208–10.

2. Merck officials also reported that Columbia was seeking more than $1 million before allowing the release of the materials, a request the company rejected on the grounds that the entire project would cost $5 million a year at most. Merck wanted to spend all its funds to support the work to be done at Washington University. However, Michael Crow, Columbia's vice provost, does not recall that Columbia asked for any such payment.

1: Engineers Plant a Rose

1. James D. Watson, "The Double Helix: A Personal Account of the Discovery of the Structure of DNA," *Mentor*, New York, 1968, p. 18.

2. Insulin was licensed to Eli Lilly.

3. Robert Cook-Deegan, "The Alta Summit," *Genomics*, volume 5, October 1989, pp. 661–63.

4. Robert Cook-Deegan, *The Gene Wars: Science, Politics, and the Human Genome*, W. W. Norton & Company, New York, 1994, p. 81. The chancellor's name was Robert Sinsheimer.

5. Ibid., p. 83.

6. This was the National Gene Library Project, pioneered by molecular biologist Larry Deaven.

7. Robert Cook-Deegan, *The Gene Wars: Science, Politics, and the Human Genome*, W. W. Norton & Company, New York, 1994, p. 97.

8. Robert Kanigel, "The Genome Project," *The New York Times Magazine*, December 13, 1988, p. 98.

9. Robert Cook-Deegan, *The Gene Wars: Science, Politics, and the Human Genome*, W. W. Norton & Company, New York, 1994, p. 99.

10. Robert S. Service, "Objection #3: Impossible to do," *Science*, February 16, 2001, p. 1186.

11. Robert Cook-Deegan, *The Gene Wars: Science, Politics, and the Human Genome*, W. W. Norton & Company, New York, 1994, p. 180; Bart Barrell, "DNA Sequencing: Present Limitations and Prospects for the Future," *FASEB Journal*, January 1990, pp. 40–45.

12. Jerry E. Bishop and Michael Waldholz, *Genome: The Story of Our Astonishing Attempt to Map All the Genes in the Human Body*, Simon & Schuster, New York, 1990, p. 219; Robert Cook-Deegan, *The Gene Wars: Science, Politics, and the Human Genome*, W. W. Norton & Company, New York, 1994, pp. 110–11.

13. Robert Cook-Deegan, *The Gene Wars: Science, Politics, and the Human Genome*, W. W. Norton & Company, New York, 1994, p. 111.

14. J. B. Walsh and J. Marks, "Sequencing the Human Genome," *Nature*, 322, August 14, 1986, p. 590.

15. Jerry E. Bishop and Michael Waldholz, *Genome: The Story of Our Astonishing Attempt to Map All the Genes in the Human Body*, Simon & Schuster, New York, 1990, p. 221.

16. Leslie Roberts, "Controversial from the Start," *Science*, February 16, 2001, p. 1182.

17. "Report on the Human Genome Initiative for the Office of Health and Environmental Research," prepared by the Subcommittee on Human Genome of the Health and Environmental Research Advisory Committee for the U.S. Department of Energy, Office of Energy Research, Office of Health and Environmental Research, April 1987.

18. Robert Cook-Deegan, *The Gene Wars: Science, Politics, and the Human Genome*, W. W. Norton & Company, New York, 1994, pp. 131–32; ref. American Society of Biochemistry and Molecular Biology (7 June 1987), Council Policy Statement on Mapping and Sequencing the Human Genome.

19. National Research Council, Mapping and Sequencing the Human Genome, National Academy Press, 1988, Washington, D.C.; Robert Cook-Deegan, *The Gene Wars: Science, Politics, and the Human Genome*, W. W. Norton & Company, New York, 1994, p. 132.

20. Victor A. McKusick, "Mapping and Sequencing the Human Genome," *The New England Journal of Medicine* 320: 910–15, 1989; Jerry E. Bishop and Michael Waldholz, *Genome: The Story of Our Astonishing Attempt to Map All the Genes in the Human Body*, Simon & Schuster, New York, 1990, p. 222.

21. Jerry E. Bishop and Michael Waldholz, *Genome: The Story of Our Astonishing Attempt to Map All the Genes in the Human Body*, Simon & Schuster, New York, 1990, p. 223.

22. The National Human Genome Research Institute, "Understanding Our Genetic Inheritance: The U.S. Human Genome Project. The First Five Years: Fiscal Years 1991–1995."

23. Robert Cook-Deegan, *The Gene Wars: Science, Politics, and the Human Genome*, W. W. Norton & Company, New York, 1994, p. 148; "The Human Genome Program Report," United States Department of Energy, Office of Energy Research, Office of Biological and Environmental Research, 1997.

2: A Rebel Lands a Cause

1. Watson does not remember making Venter this specific offer, but he does remember being impressed with Venter's sequencing operation and wanting to fund his sequencing work.

2. Robert Cook-Deegan, *The Gene Wars: Science, Politics, and the Human Genome*, W. W. Norton & Company, New York, 1994, p. 314.

3. Watson remembered that Venter's grant was not funded because the scientific reviewers chose to fund a sequencing project proposed by Leroy Hood instead. Hood's application would have been harder to reject, Watson felt, because Hood had invented the automated sequencing technology both he and Venter were using.

3: Gene Darling

1. Alice Wexler, *Mapping Fate: A Memoir of Family, Risk, and Genetic Research*, University of California Press, Berkeley, 1995, pp. 77–79.

2. Adapted from Nancy Wexler, "Clairvoyance and Caution: Repercussions of the Human Genome Project," in *The Code of Codes: Scientific and Social Issues in the Human Genome Project*, by D. J. Kevles and L. Hood (eds)., Harvard University Press, Cambridge, 1992, pp. 211–43.

3. The particular markers conceived of at that time were known as restriction-fragment-length polymorphisms (RFLPs or "riflips"). These normal DNA variations were identified via tiny molecular scissors known as restriction enzymes, which cut DNA at unique spots with a particular genetic code. The location of these cutting sites is different in different people, so the DNA pieces that result are different lengths in different people. Researchers have a way of detecting the fragment lengths, so they sought to link a particular pattern of lengths to the disease. Where those DNA pieces came from in the genome would then mark the general location of the disease gene. Other types of markers are more commonly used today.

4. This leads to a greater diversity in the offspring, since each child inherits any of various possible chromosome combinations from his or her parents rather than either one whole chromosome or the other.

5. Alice Wexler, *Mapping Fate: A Memoir of Family, Risk, and Genetic Research*, University of California Press, Berkeley, 1995, pp. 182–210.

6. Ibid.

7. Jerry E. Bishop and Michael Waldholz, *Genome: The Story of Our Astonishing Attempt to Map All the Genes in the Human Body*, Simon & Schuster, New York, 1990, p. 289.

8. In one test, for example, they looked to see whether the gene was used in the tissues affected by the disease: the lung, pancreas, and sweat glands. The gene would be used in these tissues if it were responsible for cystic fibrosis.

4: Crouching TIGR

1. Robert Cook-Deegan, *The Gene Wars: Science, Politics, and the Human Genome*, W. W. Norton & Company, New York, 1994, p. 181.

2. Ibid., p. 314.

3. Leslie Roberts, "Controversial from the Start," *Science*, February 16, 2001, which references *Science*, October 11, 1991, p. 184.

4. G. D. Schuler, et al., "A Gene Map of the Human Genome," *Science*, October 25, 1996. See reference 5 in that article: A protein database called Swiss-Prot contained only 1,790 human sequences as of August 19, 1991, according to A. Bairoch and R. Apweiler, *Nucleic Acids Research* 24, 21 (1996).

5. John Carey with Joan O'C. Hamilton, Julia Flynn, and Geoffrey Smith, "The Gene Kings," *Business Week*, May 8, 1995, p. 76.

5: Dragon Slayer

1. "Interview with Florence Haseltine, Ph.D., M.D." Conducted by Joyce Antler, Ph.D., for the Medical College of Pennsylvania on August 8, 1977, as part of the Oral History Project on Women in Medicine, and my interview with William R. Haseltine, her father, on February 27, 2002.

2. Ibid.

3. Duesberg is the retrovirologist who became famous for doubting that HIV caused AIDS.

6: The Odd Couple

1. Haseltine says that the fact that the two men would eventually terminate their association was built into their original agreements.

2. John Carey with Joan O'C. Hamilton, Julia Flynn, and Geoffrey Smith, "The Gene Kings," *Business Week*, May 8, 1995, p. 77.

3. Haseltine later estimated that the *Nature* paper contained just a small fraction (approximately one-fifth) of the genes his team had collected by the end of 1995.

7: Fame in a Germ

1. In two weeks, Sutton also made a version of the program to assemble ESTs, which had also become too voluminous for his old alignment program. In January 1995, they used it on a database of 175,000 ESTs to assemble overlapping ESTs into longer stretches of DNA and gain an estimate of how many unique full-length genes were represented in their database. The program indicated a total of about thirty-five thousand genes. The result was published later that year in *Nature*.

8: A Worm Shows the Way

1. Martin Chalfie, "The Worm Revealed," *Nature*, December 17, 1998, p. 620.

9: The Insider

1. Robert Cook-Deegan, *The Gene Wars: Science, Politics, and the Human Genome*, W. W. Norton & Company, New York, 1994, p. 335.

2. J. Madeleine Nash, "Riding the DNA Trail: Francis Collins Leads an International Drive to Track Down All the Genes and Take Their Measure," *Time*, January 17, 1994.

3. Alice Wexler, *Mapping Fate: A Memoir of Family, Risk, and Genetic Research*, University of California Press, Berkeley, 1995, p. 258.

4. Francis Collins and David Galas, "A New Five-Year Plan for the U.S. Human Genome Program," *Science*, October 1, 1993, pp. 43–46. Collins and Galas highlighted the problem in this published plan: "There is a pressing need for clone libraries with improved stability and lower chimerism and other artifacts," they wrote.

5. Researchers also hoped that YACs would copy pieces of DNA that cosmids couldn't, thereby making the map of the human genome more complete.

6. Olson's YACs, which were considerably smaller than Cohen's, were somewhat less problematic in this respect. However, yet another problem with all YACs was that it was very difficult to extract enough DNA from them.

7. They tested every chunk, cloned using YACs, for the presence of each signpost, and collected the results in a table, which showed for each chunk which if any of the signposts were contained within it. Such a table can reveal both the order of the chunks, by showing at what points (STS signposts) they overlap and, simultaneously, the order of the signposts. It worked like this: Say the library included just seven chunks—1, 2, 3, 4, 5, 6, 7—and four STS signposts A, B, C, and D. The results might read:

- Signpost A hit chunks 2, 4
- Signpost B hit chunks 2, 4, 5
- Signpost C hit chunks 1, 2
- Signpost D hit chunks 4, 5, 6, 7

So chunks 2 and 4 overlapped at signpost A; chunks 2, 4, and 5 overlapped at signpost B, and so forth. A human can solve such a simple puzzle—ordering the chunks and the signposts—by drawing lines of equal length to represent the chunks and lining them up so they overlap at the specified signposts. In this case, if you try this, you will find that the order of the signposts must be C, A, B, D and that of the chunks must be either 1, 2, 4, 5, 6, 7 or 1, 2, 4, 5, 7, 6. (Since chunk 3 has no hits on it, it cannot be ordered with respect to the others.)

8. Laura Johannes, "Detailed Map of the Genome Is Now Ready," *The Wall Street Journal*, December 22, 1995.

9. Lander's group's map, published in 1995, helped Cohen's team better anchor its chunks to the genome and thus greatly improve its map. This signpost map was useful in other ways, too, but it did not bring the start of serious human sequencing significantly closer.

10: Of Worms and Men

1. John Sulston and Georgina Ferry, *The Common Thread: A Story of Science, Politics, Ethics and the Human Genome*, Bantam Press, London, 2002, p. 118.

2. Eliot Marshall, "A Strategy for Sequencing the Genome 5 Years Early," *Science*, February 10, 1995, pp. 783–84.

3. John Sulston and Georgina Ferry, *The Common Thread: A Story of Science, Politics, Ethics and the Human Genome*, Bantam Press, London, 2002, p. 131. Translation to dollars is approximate. U.S. currency figures were calculated at the exchange rate of £1 = $1.50.

4. Eliot Marshall and Elizabeth Pennisi, "NIH Launches the Final Push to Sequence the Genome," *Science*, volume 272, April 12, 1996.

5. Ibid.

6. Christopher Anderson, "Genome Project Goes Commercial," *Science*, January 15, 1993, pp. 300–302.

11: Affliction in Iceland

1. Michael Specter, "Decoding Iceland," *The New Yorker*, January 18, 1999, p. 45.
2. Ibid., p. 45.
3. Stephen D. Moore, "Missing Link: If This Man Is Right, Medicine's Future Lies in Iceland's Past," *The Wall Street Journal Europe*, July 3, 1997.

12: The Divorce

1. John Carey with Joan O'C. Hamilton, Julia Flynn, and Geoffrey Smith, "The Gene Kings," *Business Week*, May 8, 1995.
2. On October 25, 1996, researchers from close to one hundred laboratories published the complete sequence of yeast's sixteen chromosomes and nearly six thousand genes.

13: The Man from Buffalo

1. Eliot Marshall, "Whose Genome Is It, Anyway?" *Science*, September 27, 1996, p. 1788.
2. The National Human Genome Research Institute, "Understanding Our Genetic Inheritance: The U.S. Human Genome Project. The First Five Years: Fiscal Years 1991–1995."
3. De Jong had yet to make the library from the female donor's DNA. A male donor was chosen first because men have a chromosome that females don't— the Y—and researchers wanted to get a sequence for every unique chromosome.

14: Seeds of Thunder

1. Hunkapiller recalled bringing up this idea in a conversation with White weeks earlier and that he and White told Afeyan to broach the subject at the November meeting. However, neither White nor Afeyan remembered this. The story told here is the one most consistent with the reports of the various people I interviewed.

17: Confronting the Government.

1. For instance, MIT's Whitehead Institute expected to have twenty-three million bases done by May 1998, but had completed just 8.7 million, and the Baylor group aimed for fifteen million bases but had done 8.2 million. Elizabeth Pennisi, "DNA Sequencers' Trial by Fire," *Science*, May 8, 1998, p. 814.

2. Varmus does not recall his gestures, but he does remember the encounter as tense.

3. As reported by Tony White, who heard Whitfield's talk. Whitfield could not be reached for comment.

4. John Carey, "The Duo Jolting the Gene Business: Craig Venter and Perkin-Elmer Target the Human Genome," *Business Week*, May 25, 1998.

18: The Response

1. Eliot Marshall, "NIH to Produce a 'Working Draft' of the Genome by 2001," *Science*, September 18, 1998, pp. 1174–75.

2. John Sulston and Georgina Ferry, *The Common Thread: A Story of Science, Politics, Ethics and the Human Genome*, Bantam Press, London, 2002, pp. 192–93. The cancer was successfully removed in April.

3. Incyte later decided against it.

4. Robert Langreth, "Gene-Sequencing Race Between U.S. and Private Researchers Is Accelerating," *The Wall Street Journal*, March 16, 1999.

5. The Sanger Centre was given some of the funds that had been slated for future years in advance. Translation to dollars is approximate. U.S. currency figures were calculated at the exchange rate of £1 = $1.50.

19: Demonizing a Database

1. Stephen D. Moore, "Missing Link: If This Man Is Right, Medicine's Future Lies in Iceland's Past," *The Wall Street Journal Europe*, July 3, 1997.

2. Greely was on *Nightline*, May 13, 1999. He explained that he did not distinguish between the different types of data because he didn't have time to do so on television. In speeches and documents that do not require such succinctness, Greely said he is careful to note the difference between clinical data and genetic data and that the Icelandic law applies only to the clinical data.

3. Stephen D. Moore, "Roche and Decode Announce Location of Osteoarthritis Gene," *The Wall Street Journal*, March 26, 1999.

20: The Making of a Monster

1. The stock has since split. To adjust for today's price, these figures should be halved.

2. White did not remember the particular article that prompted this interchange with Venter, but he did say that the article appeared in *The New York Times* and referred to a time frame consistent with the publication of the May 18 article.

3. This was the Genome Sequencing and Analysis Conference (GSAC).

4. Elizabeth Pennisi, "Ideas Fly at Gene-Finding Jamboree," *Science*, March 24, 2000, p. 2182.

5. Ibid., p. 2184.

6. Elizabeth Pennisi, "Fruit Fly Genome Yields Data and a Validation," *Science*, February 25, 2000, p. 1374.

7. Elizabeth Pennisi, "Ideas Fly at Gene-Finding Jamboree," *Science*, March 24, 2000, p. 2182.

8. To do this check, Myers incorporated in his final run of his assembler the small amount of fruit fly DNA sequence that the academic teams had collected. To his delight, the assembler put those pieces of sequence in the same places to which they had been pinned on the map.

21: Medicine Man

1. Nicholas Wade, "Newfound Protein Touches Off Race for New Therapies," *The New York Times*, October 31, 2000.

2. Justin Gillis, "Md. Firm's Skin Drug Heralds Genetic Leap," *The Washington Post*, September 13, 2000.

22: The Race

1. Statistically, fourfold average coverage should hit 98 percent of all bases in a chunk, but coverage of the entire genome was projected to be less, given that some chunks would be missed entirely the first time around.

2. Justin Gillis, "A Gene Dream," *The Washington Post*, September 27, 1999.

23: Hope, Hype, and Acrimony

1. Justin Gillis, "Gene-Mapping Controversy Escalates," *The Washington Post*, March 7, 2000.

2. Nicholas Wade, "Genome Decoding Plan Is Derailed by Conflicts," *The New York Times*, March 9, 2000.

3. Robert Langreth and Bob Davis, "Press Briefing Set Off Rout in Biotech," *The Wall Street Journal*, March 16, 2000.

24: *The Final Stretch*

1. Frederic Golden and Michael D. Lemonick, "The Race Is Over," *Time*, July 3, 2000.

25: *A Genome Unveiled*

1. Elizabeth Pennisi, "Finally, the Book of Life and Instructions for Navigating It," *Science*, June 30, 2000, p. 2304, and Elizabeth Pennisi, "Rival Genome Sequencers Celebrate Milestone Together," *Science*, June 30, 2000, p. 2294.

2. Nicholas Wade, "Genetic Code of Human Life Is Cracked by Scientists," *The New York Times*, June 27, 2000.

3. Frederic Golden and Michael D. Lemonick, "The Race Is Over," *Time*, July 3, 2000.

4. Scott Hensley, Laura Johannes, Rhonda L. Rundle, Thomas M. Burton, and Stephen D. Moore, "Next Milestone in Human Genetics—New DNA Map Is Expected to Advance Experiments in Gene-Based Medicine," *The Wall Street Journal*, May 26, 2000.

5. John Carey, "The Genome Gold Rush: Who Will Be the First to Hit Pay Dirt," *Business Week*, June 12, 2000.

6. Eliot Marshall, "Storm Erupts over Terms for Publishing Celera's Sequence," *Science*, December 15, 2000, pp. 2042–43.

26: *Jockeying for the Prize*

1. Myers's compartmentalized assembly program, which subdivided the shotgun data into about four thousand compartments using the public's mapping data, produced an even better genome. This one contained fifty-three thousand scaffolds, which covered 2.9 billion letters of genetic code.

2. The population of genes in an area correlated with a chemical feature. Areas packed with genes were the favored homes of just two of DNA's four chemical bases: guanine (G) and cytosine (C). Clusters of C's and G's tended to mark the beginnings and ends of genes. Scientists could see evidence of this correlation under a microscope: Because of genes' cohabitation with C-G combinations, genetic urban areas appeared as light stripes on the chromosomes. By contrast, adenines (A's) and thymines (T's) tended to hang out in the genetic deserts. These gene-dry regions appeared as dark chromosome bands.

3. Gilbert's estimate for the average span of a gene takes into account the fact that genes are themselves broken up, with the protein-coding pieces separated by intervening nonsense DNA sequences. He estimated that only 5 to 10 percent of the thirty thousand bases spanned by an average gene were actually used to make protein. In this way, Gilbert accounted for the vast bulk of junk DNA in his calculation.

4. Nicholas Wade, "Scientists Cast Bets on Human Genes; A Winner Will Be Picked in 2003," *The New York Times*, May 23, 2000; Elizabeth Pennisi, "And the Gene Number Is . . ." *Science*, May 19, 2000, p. 1146.

5. Nicholas Wade, "Human Genome Appears More Complicated," *The New York Times*, August 24, 2001.

6. Nicholas Wade, "Genome's Riddle: Few Genes, Much Complexity," *The New York Times*, February 13, 2001.

7. Ramana V. Davuluri, Ivo Grosse, and Michael Q. Zhang, "Computational Identification of Promoters and First Exon in the Human Genome," *Nature Genetics*, December 2001.

8. Paul Recer, "Researchers Charge That Celera Depended Heavily on Data from Public Research to Assemble Human Gene Structure," *Associated Press*, March 4, 2002.

9. Scott Hensley and Antonio Regalado, "Scientists Publish Critique of Celera's Work—Rivals Charge Firm Recycled Public Data in Genome Map," *The Wall Street Journal*, March 5, 2002.

10. Eric D. Green and Aravinda Chakravarti, "The Human Genome Sequence Expedition: Views from 'Base Camp,'" *Genome Research*, September 2001.

11. Scott Hensley, "Celera's Genome Anchors It Atop Biotech," *The Wall Street Journal*, February 12, 2001.

Epilogue

1. Robert F. Service, "Can Celera Do It Again?" *Science*, March 24, 2000, p. 2136.

2. Robert Langreth, "Gene Jockeys: Decoding the Human Genome in Record Time Was Easy for Celera. Producing Drugs—and Profits—Will Be Harder," *Forbes*, July 23, 2001.

3. Andrew Pollack, "Scientist Quits the Company He Led in Quest for Genome," *The New York Times*, January 23, 2002.

4. In December 2001, HGS reported disappointing preliminary results from clinical trials of its wound-healing compound Repifermin for treating internal wounds of the mouth and gastrointestinal tract. The compound proved safe at the doses used but had no discernible effect on reducing the incidence or severity of mucositis in cancer patients or, in a separate trial, in treating patients with active ulcerative colitis.

5. On April 15, 2002, HGS said it was dropping Mirostipen because the trial did not indicate that it worked well enough to warrant further development. The company said it would focus its efforts on a long-acting version of an existing medication for the same condition.

6. Scott Hensley, "HGS Directs Protein Technology at Hepatitis C Treatment," *The Wall Street Journal*, March 23, 2001.

7. Dennis Normille and Yang Jianxiang, "Japan and China Gear Up for 'Postgenome' Research," *Science*, October 5, 2001.

8. Ibid.; Li Hui, "China, Denmark Team Up to Tackle the Pig," *Science*, November 3, 2000.

9. Nicholas Wade, "Scientist Reveals Secret of Genome: It's His," *The New York Times*, April 27, 2002.

Acknowledgments

I could not have written this book without the help of numerous people, including the many eminent scientists and business people who play important roles in this story and took valuable time out of their incredibly packed schedules to speak and correspond with me. I would like to thank Hadi Abderrahim, Mark Adams, Reid Adler, Noubar Afeyan, Bruce Alberts, Bob Beliveau, Frederick Blattner, Lawrence Brody, Heidi Brumbaugh, Jim Cavanaugh, Asif Chinwalla, Sandra Clifton, John Coffin, Daniel Cohen, Stanley Cohen, Fletcher Collins, Francis Collins, Margaret Collins, David Cox, Michael Crow, Larry Deaven, Pieter de Jong, Charles DeLisi, Mitchell Drumm, Keith Elliston, Robert Fleischmann, Claire Fraser, Robert Gallo, Richard Gibbs, Walter Gilbert, Hank Greely, Sigmundur Gudbjarnason, Jeffrey Gulcher, William A. Haseltine, William R. Haseltine, Pétur Hauksson, David Haussler, Bernadine Healy, Paul Herrling, Leroy Hood, Michael Hunkapiller, Barbara Jasny, Doug Johnson, Olli-P. Kallioniemi, Donald Kennedy, James Kent, Mary-Claire King, Jonathan Knowles, Thordur Kristjansson, Raju Kucherlapati, Eric Lander, Neal Lane, Pierre Le Ber, Arnold Levine, Klaus Lindpaintner, Stephen Lombardi, Elaine Mardis, Victor McKusick, John McPherson, Paul Meltzer, Eugene Myers, Ronald Nadel, Ólafur Ólafsson, Maynard Olson, Daníel Óskarsson, Philip Ozersky, Aristides Patrinos, George Poste, Craig Rosen, Leon Rosenberg,

Gerald Rubin, Steven Scherer, Greg Schuler, Melvin Simon, Friðrik Skúlason, Hamilton Smith, Joseph Sodroski, Dan Soppett, Kári Stefánsson, John Sulston, Granger Sutton, Thorgeir Thorgeirsson, Karl Tryggvason, Lap-Chee Tsui, Harold Varmus, Craig Venter, Bert Vogelstein, Alan Walton, Robert Waterston, James Watson, James Weber, Tony White, Alan Williamson, James Wilson, Richard Wilson, Emily Winn-Deen, Barbara Wold, and James Wyngaarden.

I would also like to extend my appreciation to Stephen Fodor and Alejandro Zaffaroni. Their company Affymetrix, builder of gene chips, served as my initial inspiration for the book. I am sorry that their stories did not appear in the final version.

In writing this book, my goal was to convey the importance, substance, and excitement of an historic scientific undertaking to a lay audience. Unfortunately, that goal was not consistent with providing adequate scientific credit to all of the scientists who contributed to this revolution in science. To do so would have made the book too long. I apologize to anyone whom I had to leave out.

I'd also like to thank my agent, Al Zuckerman, and the acquiring editor, John Michel, both of whom saw the potential of the book when few others did. My editor at Times Books, Heather Rodino, provided numerous suggestions that greatly enhanced the book's readability. I appreciated her insight, kindness, and enthusiasm throughout.

My family's and friends' help and support was also touching and invaluable. My mother, Barbara Gordon-Lickey, flew across the country to care for my infant son, Lars, as I was completing the book's first draft. I also am indebted to Ellen Langreth, Katrina and John Purisima, Xiao-qian "Tina" Chen, and Suzanne and Bill Nicholas for taking time out of their busy lives to be with my son so I could write. Lars himself deserves recognition for cheerfully playing with all these wonderful people while his mommy worked.

But my deepest gratitude goes to my husband, Robert Langreth, who sacrificed countless hours to ensure that this book not only got done but that it told an accurate and compelling story. He read and edited my manuscript at many stages of completion, suggested strategies for reaching out to reluctant sources and for organizing chapters, and offered endless encouragement and enthusiasm. He helped make the book what it is.

Index

About the Author

INGRID WICKELGREN is a science journalist and contributing correspondent for *Science* magazine. The author of two previous books, she has written for various publications including *The New York Times, Business Week, Health, Discover, Science News,* and *Popular Science.* She lives with her husband and son in New Jersey.

5 7/05